Test Bank

to accompany

Ecology and Field Biology

Fifth Edition

Robert Leo Smith

Prepared by

Terril Long
Capital University

An Imprint of Addison Wesley Longman, Inc.

Menlo Park, California • Reading, Massachusetts • New York • Harlow, England
Don Mills, Ontario • Sydney • Mexico City • Madrid • Amsterdam
Bonn • Paris • Milan • Singapore • Tokyo • Seoul • Taipei

ISBN 0-065-01897-4

1 2 3 4 5 6 7 8 9 10–CRS–00 99 98 97 96

The Benjamin/Cummings Publishing Company, Inc.
2725 Sand Hill Road
Menlo Park, California 94025

Contents

CHAPTER 1   Ecology: Its Meaning and Scope

Multiple-Choice

Choose the one alternative that best completes the statement or answers the question.

Pages: 5
1. Two botanists critical of the organismic approach were
    A) Lotka and Volterra.
    B) Clements and Cowles.
    C) Liebig and Shelford.
    D) Gleason and Tansley.
    Answer: D

Pages: 9
2. In the early development of ecology, controversy arose between different beliefs in
    A) the different spellings of ecology.
    B) plant and animal ecologists.
    C) organismal and individualist approaches to ecology.
    D) all of the above.
    Answer: D

Pages: 7
3. Theoretical ecology is an attempt to
    A) explain succession.
    B) cope with large amounts of unexplained ecological data.
    C) study the law of tolerance.
    D) use mathematical models in studies of interactions and population growth.
    Answer: D

Pages: 10
4. According to H. Odum,
    A) all things are present if we just look.
    B) the natural assemblage of plants and animals of an ecosystem has been determined by
       competitive abilities and other attributes of the component species.
    C) population studies such as mutualism, competition, and predation produce mathematically
       precise equations.
    D) all is chaos in ecology.
    Answer: B

Pages: 10
5. Ecology became an applied field when
    A) the role of fire in control of plant succession was proposed in the book The Bobwhite Quail
       by Stodard in 1932.
    B) Lotka and Volterra developed mathematical methods.
    C) Clements described succession.
    D) Darwin proposed evolutionary theory.
    Answer: A

Pages: 11

6. Rachel Carson
   A) did more than any other to bring ecology to the public eye.
   B) wrote the book <u>Silent Spring.</u>
   C) was the first to realize carnivorous birds would show the effects of DDT first.
   D) All of the above
   Answer: D

Pages: 10

7. Which of the following provide(s) (an) example(s) of a major U.S. ecodisaster?
   A) the Dust Bowl
   B) the Rose Bowl
   C) DDT
   D) A and C
   Answer: D

Pages: 9

8. Two fragments or divisions in current ecological endeavors are
   A) plants and fungi.
   B) plants and animals.
   C) holistic ecosystem and reductionist evolutionary and population ecology.
   D) natural selection and evolution.
   Answer: C

Pages: 9

9. The "organismic levels of hierarchy" concept
   A) was advanced by Clements and Wheeler.
   B) is all modern ecology.
   C) was originally proposed by R.L. Smith in the early 1800s.
   D) is not an ecological concept.
   Answer: A

Pages: 11

10. Public awareness of ecological concepts is observed in
    A) extensive bulldozing of trees at a housing development.
    B) the founding of The Nature Conservancy.
    C) construction of more large killer dams to provide power for western states rather than
       developing alternatives.
    D) allowing clearcut logging of all trees over one inch in what remains of the ancient forests
       of the Northwest and in the East.
    Answer: B

Pages: 12

11. In the future, problems in foreign countries should include
    A) toxic wastes in ground waters.
    B) air pollution.
    C) deforestation.
    D) all of the above.
    Answer: D

Pages: 6

12. The newer field of sociobiology developed in the book <u>Sociobiology: The New Synthesis</u> by E.O.Wilson considers sociobiology as a study of
    A) plants in groups.
    B) evolution applied to social behavior in animals.
    C) social behavior of solitary animals.
    D) all of the above.
    Answer: B

Pages: 9

13. The book <u>Animal Communities in Temperate America</u> in 1913 was a major effort because the author stressed the relationship of plants and animals and emphasized the science of communities. The author was
    A) C. Darwin
    B) V. Shelford
    C) K. von Frisch
    D) A.E. Tansley
    Answer: B

Pages: 6

14. The close relationship between the environment and animals was proposed in the book <u>Animal Ecology.</u> The author was also instrumental in founding The Nature Conservancy. He was
    A) H. Odum
    B) C. Elton
    C) R. Carson
    D) C. Darwin
    Answer: B

Pages: 8

15. _____ developed an ecological approach to freshwater biology based on consumers and producers which was a major idea in its time.
    A) J.M. Coulter
    B) J. Warming
    C) F.E. Clements
    D) A. Thienemann
    Answer: D

Pages: 8

16. Although not an ecologist, _____ in his landmark paper "The Accumulation of Energy in Plants" which appeared in the "Ohio Journal of Science" in 1926 marked the beginning of energy budgets.
    A) F.E. Clements
    B) A. Kerner
    C) C. Darwin
    D) E. Transeau
    Answer: D

Pages: 13

17. In the development of ecological concepts, the most probable sequence of events is
    A) social ecology, behavioral ecology, succession.
    B) limnology, plant geography, succession.
    C) plant geography, succession, population ecology, ecosystem ecology.
    D) behavioral ecology, social ecology, climax concept.
    Answer: C

Pages: 4
18. The areas of historical development are
    A) plant studies and energy flow, population studies, animal studies and cooperative studies.
    B) plant studies and animal studies.
    C) behavioral and social studies.
    D) energy flow and animal studies.
    Answer: B

Pages: 9
19. "One of the fruitful ideas contributed by biological science to modern civilization" was a statement made by Allee et. al. about
    A) succession.
    B) the individualistic concept.
    C) the limnology concept.
    D) the organismic concept.
    Answer: D

Pages: 9
20. In the field of ecology there is/are
    A) peaceful cooperation between groups.
    B) noisy coexistence and agreements between various groups.
    C) strong debates and disputes between groups.
    D) heavenly bliss.
    Answer: C

True-False

Write T if the statement is true and F if the statement is false.

Pages: 4
1. Ecology is the study of the relationship between organisms and environment.
   Answer: True

Pages: 6
2. Social biology studies animal behavior.
   Answer: False

Pages: 6
3. Ecological studies are always conducted in the natural environment and never in the laboratory.
   Answer: False

Pages: 5
4. Succession is an important concept limited to animal ecology.
   Answer: False

Pages: 8
5. Limnology is the study of freshwater ecology.
   Answer: True

Pages: 6
6. Ethology is the study of function and evolution of behavior patterns.
   Answer: True

Pages: 4
7. Early ecologists were plant geographers.
Answer: True

Pages: 8
8. Producers and consumers are present in only a few communities.
Answer: False

Pages: 8
9. The International Biology Program was initiated by Great Britain.
Answer: False

Pages: 9
10. The individualistic concept states that each community arises in a prearranged manner.
Answer: False

Short Answer

Write the word or phrase that best completes each statement or answers the question.

Pages: 4
1. _____ first coined the term ecology.
Answer: E. Haeckel

Pages: 4
2. A study of the structure and function of nature, organisms in their natural habitat, is called

_____.
Answer: ecology

Pages: 5
3. The botanist who first provided an understanding of succession in his "Plant Life of the Danube Basin" was _____.
Answer: A. Kerner

Pages: 5
4. In the U.S. a dominant ecologist, _____ gave ecology a hierarchal framework, introduced many terms such as environmental indicators, and developed the concept of succession.
Answer: F.E. Clements

Pages: 8
5. The concepts of thermal stratification, nutrient cycling, and the term limnology were developed by _____.
Answer: A. Thienemann

Pages: 8
6. An ecologist who wrote "The Lake as a Microcosm" was _____.
Answer: S.A. Forbes

Pages: 7
7. _____ proposed a theory of evolution and the origin of species based on fossil remains, observation of birds and other evidence. Although not an ecologist he was an influence on ecology.
Answer: C. Darwin

Pages: 6

8. _____ developed the "law of tolerance" and authored the book <u>Animal Communities in Temperate America</u>.
   Answer: V. Shelford

Pages: 6

9. _____ coined the phrase of the law of limiting factors in his study of nutrients in 1840.
   Answer: J. Liebig

Pages: 8

10. _____ and _____ were two ecologists known for theoretical and mathematical models.
    Answer: A. Lotka and V. Volterra   (any order)

Pages: 11

11. _____ wrote the book <u>Silent Spring,</u> which brought environmental problems and the principles of ecology to the public.
    Answer: R. Carson

Matching

Choose the item from Column 2 that best matches each item in Column 1.

Pages: 8

1. Column 1: Theoretical ecology
   Column 2: Lotka and Volterra
      Foil: A. Kerner

Pages: 9

2. Column 1: <u>Silent Spring</u>
   Column 2: R. Carson
      Foil: P. Sears

Pages: 6

3. Column 1: Animal behavior
   Column 2: K. Lorenz
      Foil: T. Long

Pages: 6

4. Column 1: Law of minimum
   Column 2: J. Liebig
      Foil: R.L. Smith

Pages: 6

5. Column 1: Law of tolerance
   Column 2: V. Shelford

Pages: 11

6. Column 1: The bible of the environmental movement
   Column 2: A. Leopold

Pages: 5

7. Column 1: Ecosystem
   Column 2: A.E. Tansley

Pages: 8
8. Column 1: Limnology
    Column 2: A. Thienemann

Pages: 11
9. Column 1: "Deserts on the March"
    Column 2: P. Sears

Pages: 5
10. Column 1: Plant succession
    Column 2: F. Clements

CHAPTER 2   Experimentation and Models

Multiple-Choice

Choose the one alternative that best completes the statement or answers the question.

Pages: 16
1. Testing the hypothesis and reaching a conclusion has become prominent in ecology since
   A) 1860
   B) 1900
   C) 1960
   D) 1995
   Answer: C

Pages: 18
2. Testing a hypothesis and reaching a conclusion requires
   A) a generalized statement about nature and natural phenomena.
   B) a null hypothesis.
   C) a data hypothesis.
   D) a series of other hypotheses about many things.
   Answer: B

Pages: 16
3. In natural experiments _____.
   A) random numbers and collected data are compared.
   B) direct observation and experimentation are employed.
   C) indirect observations only are generated.
   D) deduction alone is useful.
   Answer: B

Pages: 16
4. In the hypothetic-deductive method
   A) one collects observations then infers the hypothesis.
   B) one develops a research hypothesis which is not necessary to test.
   C) one develops a research hypothesis first and then collects data to support or refute the hypothesis.
   D) All of the above
   Answer: C

Pages: 16
5. The deductive method of laboratory experimentation
   A) is sufficient in itself for ecological studies.
   B) must be verified with field studies.
   C) is of no use in ecology.
   D) must be free of null hypotheses.
   Answer: B

Pages: 22

6. Simulation models
   A) are complex and cannot be solved without a computer.
   B) are useful with simple hand calculators.
   C) require a group of models so investigators can pick and choose.
   D) All of the above
   Answer: A

Pages: 16

7. The inductive method
   A) goes from the specific to the general.
   B) goes from the general to the specific.
   C) goes from the general to an hypothesis.
   D) All of the above
   Answer: A

sample

Pages: 16

8. The deductive method
   A) goes from the specific to the general.
   B) goes from the general to the specific.
   C) goes from the specific to an hypothesis.
   D) All of the above
   Answer: B

Pages: 16

9. A hypothesis is
   A) a statement about causal agents which can be tested.
   B) a statement about causal agents which cannot be tested.
   C) a statement about an inference but not causal.
   D) None of the above
   Answer: A

Pages: 24

10. Ecology has moved from a descriptive science to a/an
    A) empirical approach.
    B) experimental approach.
    C) modeling approach.
    D) All of the above
    Answer: D

True-False

Write T if the statement is true and F if the statement is false.

sample

Pages: 16

1. Two approaches to testing hypothesis are the inductive method and the deductive method.
   Answer: True

Pages: 18

2. If it is based on facts, a model need not be verified.
   Answer: False

3. Analytical models must be solved by computer.
Answer: False

4. Ecology today is an analytical and emperical science.
Answer: True

5. A model is an abstraction and simplification of natural phenomena.
Answer: True

Short Answer

Write the word or phrase that best completes each statement or answers the question.

1. A _____ is a statement about cause that can be confirmed by an experiment.
Answer: hypothesis

2. A statement of cause that has considerable evidence from several sources that cannot be directly confirmed is _____.
Answer: theory

3. _____ involves manipulation using a hypothesis and collecting data by direct observation or measurements.
Answer: Experimentation

4. There are two approaches to testing hypotheses _____ and _____.
Answer: direct observation and experimentation.   (any order)

5. Infering generalities from observation and correlation is the basis of the _____ method.
Answer: deductive

6. The use of mathematical models means those that cannot be solved analytically are solved using the _____.
Answer: computer

7. Models are abstractions developed to _____ new results or provide insights.
Answer: predict

8. Originally ecology was a _____ field now it is more analytical.
Answer: discriptive

9. Once developed, a model must be _____ and _____.
   Answer: validated and verified   (any order)

10. A hypothesis might be _____ if it does not fit the data.
    Answer: null

Matching

Choose the item from Column 2 that best matches each item in Column 1.

1. Column 1: In the use of models, _____ is the most important step.
   Column 2: validation
     Foil: data management

2. Column 1: Testing a hypothesis entails first collection of data by direct observation or

   _____.

   Column 2: experimentation
     Foil: theory

3. Column 1: A theory is a statement of _____.
   Column 2: cause
     Foil: hypothesis

4. Column 1: Field experiments are more realistic than _____ experiments.
   Column 2: laboratory
     Foil: null hypothesis

5. Column 1: Field experiments involve manipulation of _____.
   Column 2: independent variables
     Foil: communities

CHAPTER 3 Adaptation

Multiple-Choice

Choose the one alternative that best completes the statement or answers the question.

Pages: 30
1. The ability of a genotype to cause many phenotype expressions is
   A) genetic drift
   B) phenotype plasticity
   C) a deme
   D) the founder effect
   Answer: B

Pages: 30
2. Adaptation can refer to a _____ which is the result of natural selection.
   A) behavioral trait
   B) morphological trait
   C) physiological trait
   D) all of the above
   Answer: D

Pages: 30
3. Physiological or morphological features or forms of behavior used to explain the ability of an organism to survive where it does is called:
   A) a deme
   B) an adaptation
   C) a gene pool
   D) a genotype
   Answer: B

Pages: 30
4. A population adapted to local conditions
   A) is a genotype.
   B) has no alleles.
   C) is an ecotype.
   D) did not arise by evolutionary means.
   Answer: C

Pages: 30
5. Organisms become adapted to a specific environment
   A) over long periods of time.
   B) over many generations.
   C) because there is a selection for the fittest.
   D) all of the above
   Answer: D

6. Blackman
   A) discovered the law of tolerance.
   B) proposed the law of limiting factors.
   C) proposed a neo Darwinian theory of evolution.
   D) discovered the law of maximum factors.
   Answer: B

7. Leibig
   A) developed the law of the minimum.
   B) said minerals are required by plants. If these are present in minimal quantities, minimal growth will occur.
   C) plants do not grow well even under optimal conditions.
   D) A and B.
   Answer: D

8. In Appalachian regions the soil is thin and the bedrock contains few minerals, consequently plant growth is occasionally stunted. This is an example of
   A) the concept of adaptation.
   B) the poikilotherm plants.
   C) the concept of greatest fitness.
   D) the law of minimum.
   Answer: D

9. The law of tolerance was developed by
   A) Liebig
   B) Blackman
   C) Shelford
   D) none of the above
   Answer: C

10. Homeostasis requires
    A) negative feedback.
    B) positive feedback.
    C) receptors.
    D) all of the above
    Answer: A

Sampled
11. In applying the law of tolerance, it must be realized that
    A) tolerance values may change seasonally.
    B) optimum values may be wide.
    C) on the extremes of tolerances, organisms might be less able to survive.
    D) all of the above
    Answer: D

True-False

Write T if the statement is true and F if the statement is false.

Pages: 33

1. Homeotherms allow internal environments to follow the external environments.
Answer: False

Pages: 33

2. Poikilotherms have internal environments regulated by metabolism.
Answer: False

Pages: 33

3. Homeostasis is the maintenance of a constant internal environment in the face of a variable external environment.
Answer: True

Pages: 33

4. Homeostasis requires a set point and feed back to operate.
Answer: True

Pages: 32

5. The short term response of an organism to a change in environment is acclimatization.
Answer: True

Pages: 30

6. A frog moves into a deeper mud in the <u>winter</u> and enters a <u>dormant</u> state. This is an adaptation to <u>cold</u> conditions.
Answer: True

Pages: 32

7. When a frog moves to a warmer area and suns himself to warm his body he is acclimatizing.
Answer: False

Pages: 30

8. Hibernation in general is an adaptation to a cold or winter condition.
Answer: True

Pages: 30

9. Plants which live in ravines and are unique from other plants of the same species living around the ravine make up an ecotype.
Answer: True

Pages: 33

10. Body temperature in ecology students is not an example of homeostasis.
Answer: False

Short Answer

Write the word or phrase that best completes each statement or answers the question.

Pages: 30
1. The spines of cacti are a/an _____ to desert life.
   Answer: adaptation

Pages: 31
2. Minerals are commonly very low in some environments and plant growth is reduced. This is an example of _____.
   Answer: the law of minimum.

Pages: 33
3. Body temperature of most of the students in the class is 98.6 degrees Fahrenheit. This is because temperature regulation is a result of a mechanism called _____ which depends on _____ and a temperature called the _____.
   Answer: homeostasis, feedback, set point   (in order)

Pages: 30
4. A/An _____ is a group of individuals with genetic characteristics which limit the group to an ecological area.
   Answer: ecotype

Pages: 30
5. The characteristics of gene expression are the _____.
   Answer: phenotype

Pages: 30
6. The control of mating which excludes some phenotypes and includes others is called _____.
   Answer: natural selection

Pages: 30
7. The _____ of a species is the result of the diversity of genetic characteristics.
   Answer: adaptability

Pages: 32
8. A short term adjustment of an organism to a change in the environment is called _____.
   Answer: acclimatization

Pages: 31
9. _____ refers to the range of an environmental variable which can be tolerated by an organism and will allow it to grow.
   Answer: Tolerance

Pages: 33
10. Animals which regulate body temperature by selecting an appropriate environment are called _____.

    Answer: poikilotherms

Matching

Choose the item from Column 2 that best matches each item in Column 1.

Pages: 33
1. Column 1: A snake that basks in the sun to acquire warmth is a _____.
   Column 2: endotherm

Pages : 33
2. Column 1: "Warm blooded" animals such as birds are called _____.
   Column 2: homeotherm

Pages : 31
3. Column 1: Cold temperature in winter slows the growth of winter growing plants. This is an example of _____.
   Column 2: minimum

Pages : 30
4. Column 1: Where the winter temperatures go below -5 or -10 degrees some plants are not found. This is the principle of _____.
   Column 2: limiting factor

Pages : 33
5. Column 1: An animal whose temperature fluctuates with environmental temperature is said to be a/an _____.
   Column 2: poikilotherm

Pages : 30
6. Column 1: Fish swim to the bottom of the pond in summer where it is much cooler. This is a _____.
   Column 2: behavioral response

Pages : 30
7. Column 1: The thick fleshy stems and leaves of succulents is a/an _____ to a dry habitat.
   Column 2: adaptation

Pages : 33
8. Column 1: The regulation of temperature by regulating metabolism is a characteristic of _____.
   Column 2: endothermism

Pages : 30
9. Column 1: The wings of birds are a/an _____ for flight.
   Column 2: adaptation

Pages : 32
10. Column 1: Fish shift their tolerance to seasonal change in water temperature by _____.
    Column 2: acclimatization

Essay

Write your answer in the space provided or on a separate sheet of paper.

Pages: 31

1. Many organisms, particularly those that are widespread, are able to exist over a wide range of environmental conditions. What does this tell you about the concepts of tolerance, limiting factors, and minimum?

   Answer: Organisms with widespread distribution are highly adaptable to a wide range of environmental conditions. The range of tolerance relates to three aspects: survival, growth, and reproduction. Environmental conditions suitable for survival and growth may be minimal or unsuitable for reproduction. Thus environmental conditions, such as temperature or moisture, needed for successful reproduction can define the limits of its natural range. Minimal conditions may allow survival alone.

Pages: 30

2. How does the concept of adaptation differ from the concept of acclimatization?

   Answer: Adaptation is a long term genetic change modifying fitness and acclimatization is a short term adjustment to the environment.

Pages: 33

3. How do poikilotherms differ from homeotherms? Give examples of each.

   Answer: Homeotherms maintain relatively constant body temperature by heat of metabolism. Poikilotherms maintain body temperature by absorbing heat from the environment.

Pages: 33

4. How does homeostasis of body temperature work?

   Answer: Sensors in the body detect body temperature, signals go to the brain, then to the hypothalamus which causes shivering or increased metabolic rate.

CHAPTER 4   Climate

Multiple-Choice

Choose the one alternative that best completes the statement or answers the question.

Pages: 36
1. Climate is
   A) the general pattern of weather over several years.
   B) the weather at a given time and place.
   C) a changing condition from one day to another.
   D) not as important as weather as a limiting factor.
   Answer: A

Pages: 36
2. The incoming radiation from the sun contains
   A) visible light.
   B) infra-red waves.
   C) ultra-violet light.
   D) all of the above
   Answer: D

Pages: 36
3. Of the incoming radiation from the sun, _____ is absorbed by the atmosphere.
   A) 20%
   B) 30%
   C) 40%
   D) 50%
   Answer: D

Pages: 37
4. Albedo refers to
   A) absorbed radiation.
   B) reflected radiation.
   C) radiation absorbed by the clouds.
   D) far-red radiation absorbed by the clouds.
   Answer: B

Pages: 37
5. The atmospheric gas which absorbs incoming energy from the sun is
   A) oxygen.
   B) nitrogen.
   C) argon.
   D) carbon dioxide.
   Answer: D

Pages: 37
6. The substance in clouds which absorbs the sun's energy is
   A) oxygen.
   B) nitrogen.
   C) water.
   D) all of the above.
   Answer: C

18

Pages: 37
7. When the greenhouse effect occurs,
    A) atmospheric temperature decreases.
    B) atmospheric temperature near the ground is warmer than 2 ft. up in the air.
    C) winds are generated and the clouds are blown away.
    D) radiation reflects back to the earth warming the air.
    Answer: D

Pages: 41
8. The sun's energy is unevenly distributed on the earth's surface because
    A) the rounded shape means that less energy per unit area reaches the surface at the poles.
    B) tilting of the earth on its axis means the hemispheres are affected.
    C) the orbit is elliptical.
    D) all of the above
    Answer: D

Pages: 41
9. The average air temperature remains relatively constant because
    A) there is fluctuation of absorbed energy at the tropics.
    B) heat moves from larger latitudes to the equator.
    C) low latitudes have a greater variation in incoming radiation.
    D) there is steady radiation at the equator and warm air rises and moves to higher latitudes.
    Answer: D

Pages: 48
10. The movement of cool moist ocean air over warm air on land forms a
    A) marine inversion.
    B) radiation inversion.
    C) albedo.
    D) greenhouse effect.
    Answer: A

Pages: 39,42
11. Because of the Coriolis effect
    A) airflow in the northern hemisphere is deflected clockwise.
    B) airflow in the southern hemisphere is deflected counter-clockwise.
    C) in the northern hemisphere, cyclones flow counter-clockwise.
    D) all of the above
    Answer: D

Pages: 41
12. Cool
    A) light air rises over arctic regions, then flows in circles around the poles.
    B) heavy air rises over the tropics.
    C) heavy air sinks over the arctic regions and flows toward the equator.
    D) all of the above
    Answer: C

Pages: 41

13. The earth has _____ cells of air flow.
    A) two
    B) four
    C) six
    D) eight
    Answer: C

Pages: 42

14. Circular ocean currents
    A) occur in the Atlantic Ocean.
    B) occur in the Pacific Ocean.
    C) are called gyres.
    D) all of the above
    Answer: D

Pages: 41

15. Wind patterns in the middle of the Northern Hemisphere are
    A) easterlies.
    B) westerlies.
    C) doldrums.
    D) polar.
    Answer: B

Pages: 36

16. The rise in temperature experienced during the greenhouse effect is due to the presence of
    A) hydrogen.
    B) nitrogen.
    C) oxygen.
    D) carbon dioxide.
    Answer: D

Pages: 36

17. Air masses with high pressure are accompanied by
    A) rainy weather.
    B) cloudy weather.
    C) clear sky.
    D) thunderstorms.
    Answer: C

Pages: 40

18. The sun is directly overhead on the equator
    A) at the spring equinox.
    B) at the fall equinox.
    C) both A and B
    D) none of the above
    Answer: C

19. Dense vegetation controls the microclimate by
    A) creating turbulent air flow.
    B) absorbing most of the light.
    C) altering soil moisture by reducing evapotranspiration.
    D) all of the above
    Answer: D

20. Humans change temperatures of the atmosphere by
    A) removing the trees.
    B) building large buildings.
    C) constructing cities.
    D) all of the above
    Answer: D

21. In the northern hemisphere, which of the following is true?
    A) South facing slopes are usually cooler than north facing slopes.
    B) South facing slopes usually have more soil moisture than north facing slopes.
    C) North facing slopes are usually cooler than south facing slopes.
    D) all of the above
    Answer: C

22. Climate
    A) determines plant distributions.
    B) determines animal distributions.
    C) determines soil moisture.
    D) all of the above
    Answer: D

23. A biome found in the continental United States is the
    A) grassland.
    B) antarctic.
    C) beech maple woods.
    D) tundra.
    Answer: A

24. A microclimate could be
    A) the arctic region.
    B) the region of the deciduous forest biome.
    C) the weather over a continent.
    D) the temperature, humidity, etc. under a forest canopy ravine.
    Answer: D

25. The term <u>association</u> refers to plants in a
    A) unique ecosystem.
    B) a distinct habitat.
    C) a particular physical environment.
    D) all of the above
    Answer: D

26. Inversions are pronounced in
    A) flat areas.
    B) plains.
    C) deserts.
    D) hilly areas.
    Answer: D

27. Mountains influence climate by
    A) modifying patterns of precipitation.
    B) winds which blow down slopes on the windward side.
    C) warming the air.
    D) having less rain on the windward side.
    Answer: A

True-False

Write T if the statement is true and F if the statement is false.

1. Air rises when it is warmed by the sun.
   Answer: True

2. Because of the cold, one would expect timberline plants to be adapted to moist environments.
   Answer: False

3. As air is warmed it usually becomes drier.
   Answer: True

4. Temperature decreases with increasing altitude.
   Answer: True

5. Relative humidity increases with increasing temperature.
   Answer: False

6. The behavior of large air masses called <u>cells</u> determines our weather.
   Answer: True

Pages: 48

7. Vegetation alters soil temperature and moisture thus producing microclimates.
Answer: True

Pages: 45

8. Relative humidity increases over water and with increasing altitude.
Answer: True

Pages: 42

9. Ocean currents in the northern hemisphere move counter-clockwise.
Answer: False

Pages: 42

10. Currents in the southern hemisphere's gyre move clockwise.
Answer: False

Pages: 45

11. If the moisture content of air remains constant as it is cooled the relative humidity decreases.
Answer: False

Pages: 44

12. North facing slopes are usually drier than south facing slopes.
Answer: False

Pages: 54

13. The air temperature on a farm near a city is usually higher than the city.
Answer: False

Pages: 56

14. The seasonal distribution of rainfall determines the distribution of animals more than the average rainfall.
Answer: True

Pages: 36

15. The weather over a long period of time is an average condition called climate.
Answer: True

Pages: 36

16. Most of the sun's energy which strikes the earth is used in photosynthesis.
Answer: False

Pages: 53

17. When a temperature inversion occurs, air pollution is increased.
Answer: True

Pages: 39

18. The Coriolis effect turns the movement of air upward.
Answer: False

Pages: 49

19. When moist air blows up a mountain, the updraft side has a higher rainfall than the downdraft side.
Answer: True

Pages: 56
20. Biomes are small microclimates with unique life forms.
Answer: False

Pages: 44
21. Deserts are usually found in the rain shadow of mountains.
Answer: True

Pages: 39
22. When the earth spins, the surface moves faster at the equator than at 60 degrees latitude.
Answer: True

Short Answer

Write the word or phrase that best completes each statement or answers the question.

Pages: 36
1. The temperature, humidity, winds, and other outdoor conditions at one time and place are called the _____.
Answer: weather

Pages: 36
2. _____ is the average or summation of the weather conditions over a long period of time.
Answer: Climate

Pages: 36
3. The solar constant representing the amount of solar energy 83 km above the earth is _____ Langleys per minute.
Answer: 1.98

Pages: 37
4. Some _____ is emitted from the earth after being absorbed from the sun's radiation.
Answer: long-wave radiation (heat waves)

Pages: 36
5. _____ and water vapor are the principal gases in the atmosphere which absorb the sun's energy.
Answer: Carbon dioxide

Pages: 37
6. The _____ is due to the absorption of infra-red light from the sun, its re-radiation from the earth, and its reflection back to the earth from the clouds.
Answer: greenhouse effect

Pages: 39
7. The earth is tilted to the plane of its orbit about the sun by a value of _____ degrees.
Answer: 23 1/2

Pages: 46
8. An atmospheric condition producing an increase in temperature as elevation increases is called a/an _____.
Answer: inversion

Pages: 48

9. When cool moist air from the ocean moves over the land resulting in warm air near the ground and cooler air higher up, a _____ is formed.
   Answer: marine inversion

Pages: 39

10. In the United States, the _____ is a change in direction of north moving winds to a more east moving wind.
    Answer: Coriolis effect

Pages: 45

11. _____ refers to the moisture content of the air.
    Answer: Relative humidity

Pages: 48

12. _____ refers to the climatic conditions in small areas such as a rotting log or a ravine.
    Answer: Microclimate

Pages: 38

13. The temperature at which water in the air condenses is the _____.
    Answer: dew point

Pages: 54

14. Cities and other areas of human activity _____ the temperature of the earth and the air.
    Answer: raise

Pages: 38

15. Adiabatic expansion of the air results in _____.
    Answer: cooling

Pages: 38

16. Adiabatic compression of the air results in _____.
    Answer: heating

Pages: 42

17. Circular water movements in both the Atlantic and Pacific Oceans are called _____.
    Answer: gyres

Pages: 49

18. When the vegetation is removed the soil temperature at the surface becomes _____.
    Answer: warmer

Pages: 41

19. _____ move at the leading edge of large high pressure air masses.
    Answer: Weather fronts

Pages: 51

20. _____ deflects winds in a wooded area.
    Answer: Vegetation

Pages: 50

21. Plants growing on south facing slopes of the Appalachian hills require _____ than plants growing on north facing slopes.
    Answer: drier sites

Pages: 56
22. A large regional community of plants and animals with characteristic plant forms determined by climate is a _____.
   Answer: biome

Pages: 59
23. The _____ uses potential evapotranspiration rate, average total annual precipitation, and humidity to determine specific life zones.
   Answer: Holdridge life zone system

Pages: 59
24. The _____ concept involves a synthesis of climate types, vegetation, and soil types.
   Answer: ecoregion

Pages: 55
25. _____ has a pronounced effect on the distribution of vegetation.
   Answer: Climate

Matching

Choose the item from Column 2 that best matches each item in Column 1.

The solar radiation is dispersed as it impinges on the earth and its atmosphere as follows:

Pages: 36
1. Column 1: Total incoming solar radiation.
   Column 2: 100%
      Foil: 70%

Pages: 36
2. Column 1: Total reflected solar radiation.
   Column 2: 30%
      Foil: 12%

Pages: 36
3. Column 1: Total infrared emitted by the atmosphere.
   Column 2: 66%

Pages: 36
4. Column 1: Absorbed by the earth's surface.
   Column 2: 45%

Pages: 36
5. Column 1: Absorbed by the atmosphere.
   Column 2: 25%

Essay

Write your answer in the space provided or on a separate sheet of paper.

Pages: 36
1. Explain how the greenhouse effect occurs.
   Answer: CO_2 moisture in the atmosphere absorb outgoing infrared radiation and radiated it back to earth.

Pages: 54
2. What are some of the effects of cities on the environment surrounding them?
   Answer: Fog, increased precipitation, less sunshine, increased temperatures.

Pages: 39
3. Explain how the tilt and shape of the earth result in unequal distribution of solar energy on the earth.
   Answer: The sun's rays strike more directly at the equator than at the polar regions, so lower latitudes get more heat. The tilt of the earth causes the seasonal changes in heat distribution.

Pages: 44
4. There are differences in moisture between the west side of a mountain and the east side when moist westerly winds pass over. How does this occur and why is it so extreme?
   Answer: Moist air loses moisture going up the mountain and is lost before it gets to the lee side.

Pages: 56
5. What is a biome and how is it determined?
   Answer: A biome is a major regional ecological community of plants and animals--usually based on vegetation type. Climate regulates the distribution of biomes.

CHAPTER 5   Water Balance

Multiple-Choice

Choose the one alternative that best completes the statement or answers the question.

Pages: 64
1. A nearly constant temperature is maintained by large bodies of water because water has a
   A) low melting point.
   B) high viscosity.
   C) high surface tension.
   D) high specific heat.
   Answer: D

Pages: 75
2. Organisms can maintain water balance when
   A) water lost by an organism by evaporation exceeds water uptake.
   B) water loss is less than water taken up by an organism.
   C) absorption of water must exceed the loss of water.
   D) water loss is equal to water gain.
   Answer: D

Pages: 75
3. Salt glands are found in
   A) desert animals.
   B) salt water birds.
   C) salt water invertebrates.
   D) fresh water birds.
   Answer: B

Pages: 75
4. Marine animals must
   A) conserve water.
   B) eliminate salts.
   C) eliminate water.
   D) A and B
   Answer: D

Pages: 76
5. A nighttime adaptation(s) of the African ungulate, the oryx is/are
   A) large increases in body temperature.
   B) large reductions in cutaneous evaporation.
   C) large reductions in metabolic rate.
   D) pants only at very high temperature.
   Answer: B

Pages: 75
6. The kidney which eliminates salts in marine animals also achieves the same importance in _____ conservation in desert and arid mammals.
   A) water
   B) sodium
   C) chloride
   D) urea
   Answer: A

Pages: 70
7. Transpiration in a drought period does not occur
   A) through the cuticle.
   B) through open stomata.
   C) through the stem.
   D) through the upper leaf surfaces.
   Answer: B

Pages: 72
8. Reduced or thickened leaves and thick fleshy stems are adaptations to
   A) cold conditions.
   B) wet conditions.
   C) dry conditions.
   D) low altitudes.
   Answer: C

Pages: 76
9. To conserve water desert rodents
   A) have large long loops of Henle.
   B) stay in sealed burrows by day.
   C) have no sweat glands.
   D) all of the above.
   Answer: D

Pages: 68
10. Capillary water
    A) is too tightly bound to soil particles to be used by plants.
    B) will be lost due to gravitational pull when the rain stops.
    C) is not a part of field capacity.
    D) can be used by plants.
    Answer: D

Pages: 66
11. In the fresh water cycle, the largest amount of available water is found in the
    A) soil.
    B) atmosphere.
    C) lakes.
    D) rivers.
    Answer: C

12. Of the mean annual global precipitation _____ % is precipitated on land.
    A) 23
    B) 90
    C) 10
    D) 50
    Answer: A

13. In undisturbed forest soils
    A) infiltration rates are negligible.
    B) infiltration rates are greater than rainfall.
    C) surface runoff is extensive.
    D) infiltration is greater in urban areas.
    Answer: B

14. The portion of water held between soil particles is
    A) transpirational water.
    B) field capacity.
    C) hydroscopic water.
    D) capillary water.
    Answer: D

15. Poikilohydric organisms
    A) restrict growth to moist periods.
    B) grow best in dry weather.
    C) grow well when roots are underwater.
    D) maintain a stable water balance independently of atmospheric moisture.
    Answer: A

True-False

Write T if the statement is true and F if the statement is false.

1. Because of hydrogen bonding by water, water has a high specific heat and heats up rapidly in the spring.
   Answer: False

2. Water cools rapidly in the fall because of its nature.
   Answer: False

3. Ice is less dense than liquid water near freezing and as a result it floats.
   Answer: True

4. Surface tension makes it possible for water striders to walk on water.
   Answer: True

Pages: 66

5. Rain results from cooling of moist air by air masses.
Answer: True

Pages: 67

6. Interception is the driving force of the water cycle.
Answer: False

Pages: 68

7. Field capacity refers to the soil water left after the gravitational water has drained.
Answer: True

Pages: 68

8. Evapotranspiration is the loss of evaporating water from plant surfaces and the ground.
Answer: True

Pages: 72

9. Cacti lack leaves as an adaptation to a desert environment and as a result they do not carry on photosynthesis.
Answer: False

Pages: 68

10. Lakes are the prime movers of water over the globe.
Answer: False

Pages: 76

11. Since it has no water, a desert is not similar to a saline environment.
Answer: False

Pages: 76

12. Some plants and animals survive periods of drought by becoming dormant.
Answer: True

Pages: 76

13. Some desert birds like salt water birds use a salt gland to maintain water balance.
Answer: True

Pages: 74

14. Wetland plants have no adaptations for flooding, aerenchyma tissues are for salts.
Answer: False

Pages: 72

15. Plants with low concentrations of ions in their cells are halophytes.
Answer: False

Short Answer

Write the word or phrase that best completes each statement or answers the question.

Pages: 64

1. Without the cycling of _____, nutrients such as calcium, phosphorus, and sulfur could not make their endless odyssey through the ecosystem.
Answer: water

Pages: 64
2. Two hydrogen atoms are bonded to an oxygen atom by _____ bonds.
   Answer: covalent

Pages: 65
3. The water molecule is relatively _____ on the oxygen side and relatively _____ on the hydrogen side.
   Answer: negative, positive   (in order)

Pages: 64
4. The angle between the hydrogen atoms in water is _____ degrees.
   Answer: 105

Pages: 64
5. Water molecules are attracted to other water molecules (cohesion) and some different molecules (adhesion) by _____ bonds between the hydrogen in the water molecule and the oxygen in another molecule.
   Answer: hydrogen

Pages: 64
6. The vaporization of ice is called _____.
   Answer: sublimation

Pages: 64
7. To convert one gram of ice at one degree Celsius to liquid uses _____ calories of heat.
   Answer: 80

Pages: 65
8. Due to the hydrogen bonds, water shows a high resistance to flow called _____.
   Answer: viscosity

Pages: 65
9. The _____ of water is the amount of heat required to raise the temperature one degree Celsius.
   Answer: specific heat

Pages: 65
10. The attraction of water molecules to other water molecules but not to the air above the water results in a phenomena called _____.
    Answer: surface tension

Pages: 65
11. The conversion of a liquid to a gas is called _____.
    Answer: evaporation

Pages: 66
12. About _____ % of the the earth's water is found in the ocean.
    Answer: 93-97

Pages: 67
13. Precipitation of water moves into the soil by _____ and will _____ downward.
    Answer: infiltration, percolate   (in order)

Pages: 68

14. _____ is the maximum water that soil can hold after the loss of gravitational water.
    Answer: Field capacity

Pages: 68

15. The evaporation of water from the internal tissues of leaves, stems, and other living parts is called _____.
    Answer: transpiration

Pages: 69

16. _____ is the driving force of the water cycle.
    Answer: Precipitation

Pages: 70

17. The movement of water through a differentially permeable membrane from a region of higher to a region of lower concentration is _____.
    Answer: osmosis

Pages: 70

18. Plants with long tap roots that allow use of water deep in the soil are called _____.
    Answer: phreatophytes

Pages: 70

19. Plants with fleshy thick watery leaves for water storage are called _____.
    Answer: succulents

Pages: 74

20. In some species of plants, adaptations occur to constant flooding called knees (pheumatophores) which are composed of _____ tissue.
    Answer: aerenchyma

Pages: 75

21. Animals living in sea water such as bony fish have cells with lower salt content than the ocean and pump ions out of their bodies by _____.
    Answer: active transport

Matching

Choose the item from Column 2 that best matches each item in Column 1.

Pages: 64

1. Column 1: Responsible for cooling the human body in the summer.
   Column 2: Heat of vaporization

Pages: 65

2. Column 1: Allows some water insects like the water spider to walk on water.
   Column 2: Surface tension

Pages: 65

3. Column 1: Important in temperature regulation in the ocean and lakes.
   Column 2: Specific heat

Pages: 65
4. Column 1: Slows down animal movement and demands considerable energy for animal movement in water.
   Column 2: Viscosity

Pages: 65
5. Column 1: Heat required to melt one gram of ice.
   Column 2: Heat of fusion

Essay

Write your answer in the space provided or on a separate sheet of paper.

Pages: 64
1. How do the chemical properties of water explain the characteristics observed in the field?
   Answer: Water owes its unique structure to the nature of the hydrogen bonding between, and the polarity of, its molecules. Resulting in a tetrahedral arrangement. At freezing temperatures, the lattice arrangement has wide open spaces; ice occupies more space than liquid water and, as a result, floats. At 4° C water reaches its greatest density. The higher the temperature, the less stable the bonding; water becomes liquid and, at even higher temperatures, gas. The behavior of water molecules relates to formation of fog, specific heat, viscosity, capillary action, and surface tension.

Pages: 72
2. What is the water cycle?
   Answer: The water cycle starts with rain, then runoff, soil, rivers, ocean, and ends with evaporation.

Pages: 72
3. What are some mechanisms of drought resistance in plants?
   Answer: Thick cuticle, dormancy, closing stomata, deep or spreading roots, etc.

Pages: 75
4. What are some animal adaptations to drought?
   Answer: Reduction in metabolism, excretion of salt, dormancy, etc.

Pages: 76
5. What are some annual adaptations to salt in the environment?
   Answer: Salt glands, long loops of Henle, other desert adaptations.

CHAPTER 6   Thermal Balance

Multiple-Choice

Choose the one alternative that best completes the statement or answers the question.

Pages: 80
1. The diffusion of heat between two touching solid bodies with different temperatures is
   A) evaporation.
   B) radiation.
   C) conduction.
   D) convection.
   Answer: C

Pages: 86
2. Poikilotherms
   A) regulate temperature internally.
   B) are endogenous regulates of temperature.
   C) are all birds.
   D) have body temperatures regulated by the external environment.
   Answer: D

Pages: 100
3. Heat loss must _____ heat gained.
   A) equal heat stored in fat and
   B) equal
   C) equal heat gained from sweating and
   D) be less than
   Answer: B

Pages: 86
4. Metabolism _____ animals.
   A) supplies heat in
   B) uses heat from
   C) stores heat in
   D) redistributes heat in
   Answer: A

Pages: 86
5. Homeotherms are
   A) cold blooded.
   B) ectothermic.
   C) endothermic.
   D) like snakes.
   Answer: C

Pages: 86
6. In a poikilotherm, an increase in temperature will _____ the rates of enzymatic activity.
   A) increase
   B) decrease
   C) have no effect on
   D) increase for some enzymes and decrease for others
   Answer: A

Pages: 91
7. Which of the following are temperature-controlling in homeotherms?
   A) panting
   B) sweating
   C) shivering
   D) all of the above
   Answer: D

Pages: 83
8. In grasses temperature is a major factor in the distribution of species because
   A) C3 plant grasses are found in more northern regions.
   B) C3 plants are found in southern regions.
   C) distribution of grasses are not due to photosynthesis.
   D) C4 plants are found in northern regions.
   Answer: A

Pages: 94
9. A dormancy similar to hibernation but employed by desert animals is
   A) ectothermy.
   B) endothermy.
   C) estivation.
   D) thermothermy.
   Answer: C

Pages: 91
10. Endotherms are better equipped to survive than ectotherms
    A) on the equator.
    B) in a hot cave.
    C) in cold climates.
    D) in a greenhouse.
    Answer: C

Pages: 91,94
11. A small desert rat
    A) has a high metabolic rate.
    B) has a low metabolic rate.
    C) is a poikilotherm.
    D) requires less food per body weight than a bear.
    Answer: A

Pages: 92
12. Which of the following are involved in getting rid of excess heat?
    A) stored fat
    B) white feathers
    C) huge body like a bear
    D) long ears
    Answer: D

Pages: 87

13. Aquatic poikilotherms like fish have temperature zones such as
    A) the zone of thermal tolerance.
    B) the zone of thermal death.
    C) the zone of thermal resistance.
    D) all of the above
    Answer: D

Pages: 92

14. Pigs, horses, and humans can overcome a little temperature increase by
    A) running.
    B) developing nocturnal habits.
    C) increasing metabolic rate.
    D) all of the above
    Answer: B

Pages: 86

15. The amount of heat gained or lost by an animals body is related to its
    A) weight.
    B) height.
    C) volume.
    D) surface area.
    Answer: D

Pages: 89

16. Mammals insulate their bodies and slow heat loss by
    A) light colored fur.
    B) light colored feathers.
    C) subcutaneous fat.
    D) long ears.
    Answer: C

Pages: 94

17. Hibernation is
    A) a summer process.
    B) a deeper state of torpor.
    C) a plant process.
    D) limited to homoeotherms.
    Answer: B

Pages: 87

18. Many animals have a central zone of thermal tolerance like
    A) squirrels.
    B) fish.
    C) humans.
    D) salamanders.
    Answer: B

Pages: 90
19. Homeotherms maintain body temperature by
    A) ectothermy.
    B) endothermy.
    C) heterothermy.
    D) poikilothermy.
    Answer: B

Pages: 81
20. When air or water moves over an object and heat is lost, the process is:
    A) radiation.
    B) conduction.
    C) convection.
    D) homeothermic action.
    Answer: C

Pages: 87
21. Some poikilotherms are
    A) mice and squirrels.
    B) humans and birds.
    C) birds and raccoons.
    D) frogs and snakes.
    Answer: D

Pages: 86
22. As body weight increases, oxygen consumption per unit of body weight
    A) increases exponentially.
    B) decreases exponentially.
    C) decreases linearly.
    D) stays the same.
    Answer: B

Pages: 86
23. In a poikilotherm increases in body temperature result in
    A) an increase in basal metbolism.
    B) a decrease in basal metabolism.
    C) no change.
    D) sweating.
    Answer: A

Pages: 89
24. Insulation in mammals is
    A) sweat.
    B) dried grass.
    C) fur.
    D) carbohydrates.
    Answer: C

25. The metabolism of brown fat to increase body heat
    A) is called thermogenesis.
    B) is called carbon dioxide incorporation.
    C) occurs in plants.
    D) does not occur in mammals.
    Answer: A

26. Brown fat to increase body heat
    A) aids in maintaining body temperature.
    B) is found in mammals that hibernate.
    C) is found in the young of most species.
    D) all of the above
    Answer: D

True-False

Write T if the statement is true and F if the statement is false.

1. Heat transfer by convection occurs when a cool wind blows on a warm animal's body.
   Answer: True

2. Internal physiological processes maintain animal body temperature in endotherms.
   Answer: True

3. Poikilotherms do not exchange body heat with the environment.
   Answer: False

4. Ectotherms can lower temperature if the environment gets cooler.
   Answer: True

5. Endotherms depend on external physiological activities to regulate body temperature.
   Answer: False

6. Some insects can warm up by spreading their wings and orienting their bodies to the sun.
   Answer: True

7. Heterotherms regulate their body temperature to the surroundings throughout the year.
   Answer: False

8. Small animals require more food per unit volume than large animals.
   Answer: True

9. Large body size is the usual case with ectotherms.
Answer: False

10. Hibernation is a summer rest period found in some desert animals as a mechanism to survive the dry heat of summer.
Answer: False

11. If body temperatures drop a little due to heat loss, homeotherms can increase their metabolic heat production.
Answer: True

12. Bears hibernate in winter by reducing metabolic activity and increasing carbon dioxide levels.
Answer: True

13. Some hibernators feed heavily in late summer to have enouth large fat reserves for the winter.
Answer: True

14. Some common poikilotherms are field mice, squirrels, and cats.
Answer: False

15. Bumble bees loose excess heat by increasing blood flow to the central abdomen.
Answer: True

16. Plant and animal distributions are not regulated by temperature.
Answer: False

Short Answer

Write the word or phrase that best completes each statement or answers the question.

1. The movement of heat from a warm solid object to a cool one when they touch is called

_____.

Answer: conduction

2. When either air or water moves over an object and heat is moved it is said to have moved by

_____.

Answer: convection

3. When heat is released from a body in the infrared range it is said to have lost _____.
Answer: thermal radiation

Pages: 87

4. Snakes and frogs are called _____ because their body temperature is the same as the environment.
   Answer: poikilotherms

Pages: 87

5. Poikilotherms carry out their daily functions at a range of body temperatures called _____.
   Answer: the active temperature range

Pages: 82

6. _____ is the ability of aquatic poikilotherms to adjust to seasonal temperature changes.
   Answer: Acclimatization

Pages: 87

7. _____ maintain a constant body temperature despite varying environmental temperature by endogenous control.
   Answer: Homeotherms

Pages: 86

8. As body weight increases, the rate of oxygen consumption per unit of weight _____ exponentially.
   Answer: decreases

Pages: 86

9. In a poikilotherm, an increase in body temperature results in an increase in _____.
   Answer: basal metabolism

Pages: 86

10. _____ are animals like bats that sometimes regulate their body temperature and sometimes do not.
    Answer: Heterotherms

Pages: 89

11. The major form of insulation of most mammals is _____.
    Answer: fur

Pages: 94

12. By metabolizing _____ for fuel, homeotherms can increase heat production.
    Answer: brown fat

Pages: 92

13. The process of increasing the metabolism in mammals acclimated to cold temperatures is called _____.
    Answer: nonshivering thermogenesis

Pages: 95

14. Dogs and some other mammals _____ to increase evaporative cooling.
    Answer: pant

Pages: 84

15. If the onset of cold temperatures is slow, many plants can _____ by increasing sugars.
    Answer: acclimatize

Pages: 85

16. If plants acclimatize to cold weather they can be _____ for brief periods of time.
    Answer: supercooled

Pages: 85

17. Temperature wise, an animal's body can be thought of as a/an _____ and a/an _____.
    Answer: body core, outer layer   (any order)

Pages: 91

18. To maintain core temperature, the mammal must lose or gain heat by _____.
    Answer: metabolic heat production

Pages: 90

19. _____ maintain body temperature by endothermy.
    Answer: Homeotherms

Pages: 86

20. _____ maintain body temperature by ectothermy.
    Answer: Poikilotherms

Pages: 86

21. _____ utilize ectothermy and endothermy at different times.
    Answer: Heterotherms

Pages: 87

22. Many aquatic poikilotherms have a central zone of _____.
    Answer: thermal tolerance

Pages: 94

23. Hibernation is a deep state of _____.
    Answer: torpor

Pages: 88

24. Poikilotherms have a/an _____, a temperature so high that the animal cannot move to escape the condition.
    Answer: critical thermal maximum

Pages: 100

25. Temperature has a major influence on the distribution of _____.
    Answer: plants and animals

Matching

Choose the item from Column 2 that best matches each item in Column 1.

Pages: 89

1. Column 1: Birds
   Column 2: homeotherms

Pages: 86

2. Column 1: Snakes
   Column 2: poikilotherms

Pages: 87
3. Column 1: Allocate more energy to reproduction
   Column 2: poikilotherms

Pages: 89
4. Column 1: Mammals
   Column 2: homeotherms

Pages: 86
5. Column 1: Alligators
   Column 2: poikilotherms

Pages: 93
6. Column 1: Bees and butterflies
   Column 2: heterotherms

Pages: 94
7. Column 1: Birds with two daily states of torpor
   Column 2: heterotherms

Pages: 89
8. Column 1: Are endothermic
   Column 2: homeotherms

Pages: 94
9. Column 1: Bears
   Column 2: homeotherms

Pages: 86
10. Column 1: Adult insects
    Column 2: homeotherms

Pages: 88
11. Column 1: Fish
    Column 2: poikilotherms

Essay

Write your answer in the space provided or on a separate sheet of paper.

Pages: 81
1. With the breakdown of the ozone layer and burning of fossil fuels, scientists predict a global warming. What effect would this have on the earth's vegetation?
   Answer: Tropical and other plants would move northward; deciduous forest advance into conferous forest; conferous forest into tundra; tundra would shrink.

Pages: 86
2. Body temperature in poikilotherms is regulated by what mechanisms?
   Answer: Poikilotherms regulate body temperature by behavioral means by moving to cooler or warmer areas; in winter they become dormant.

Pages: 89

3. How can an endotherm loose heat when its body is cooler than the air?

   Answer: Panting, sweating, and moving to cooler areas such as burrowing in the ground.

Pages: 94

4. What happens when a heterotherm comes out of hibernation?

   Answer: Breathing and respiration increase lowering the blood carbon dioxide level, raising the blood pH; lactic acid is formed, metabolism rises.

Pages: 83

5. How does temperature control plant distribution?

   Answer: The ranges of plants, in part, are determined by thermal energy exchange. They are limited by ability (or the lack thereof) to carry on photosynthesis within certain temperature ranges, seeds require certain temperature ranges for successful germination and growth.

CHAPTER 7   Light and Biological Cycles

Multiple-Choice

Choose the one alternative that best completes the statement or answers the question.

Pages: 101
1. What factor is not important in limiting the rate of photosynthesis?
   A) quality of light
   B) amount of light
   C) moisture
   D) ultraviolet radiation
   Answer: D

Pages: 100
2. The light active in photosynthesis is
   A) red.
   B) infrared.
   C) ultraviolet.
   D) green.
   Answer: A

Pages: 100
3. Chloroplasts absorb about _____% of the PAR.
   A) 20
   B) 50
   C) 70
   D) 90
   Answer: C

Pages: 101
4. In prairie and meadow situations most of the absorption of PAR
   A) occurs at the ground level.
   B) occurs in the air above the plants.
   C) occurs in the middle and lower regions of the vegetation.
   D) does not occur.
   Answer: C

Pages: 102
5. Light which enters water
   A) is absorbed rapidly.
   B) has red absorbed rapidly.
   C) is attenuated.
   D) all of the above
   Answer: D

Pages: 103
6. Shade tolerant plants have _____ than sun tolerant plants.
   A) thinner leaves
   B) smaller leaves
   C) higher photosynthesis rates
   D) higher rates of respiration
   Answer: A

Pages: 102

7. The wavelength of light most strongly absorbed in aquatic environments is
   A) red.
   B) infrared.
   C) green.
   D) blue.
   Answer: B

Pages: 103

8. When leaves are taken from the same tree but in sunny areas and shaded areas, the sunny leaves
   are _____ than the shaded leaves.
   A) smaller
   B) thicker
   C) more deeply lobed
   D) all of the above
   Answer: D

Pages: 105

9. Destruction of the ozone layer in the stratosphere has resulted in an increase in
   A) red radiation.
   B) green radiation.
   C) yellow radiation.
   D) ultraviolet radiation.
   Answer: D

Pages: 102

10. The leaf area index is:
    A) (total leaf leaf area above a given ground)/(leaf area capable of photosynthesis)
    B) (total leaf area)/(ground area)
    C) (total area of leaves doing photosynthesis)/(total leaf area)
    D) total leaf area
    Answer: B

Pages: 106

11. Which of the following is not a circadian rhythm?
    A) The human body temperature fluctuates a few degrees being the highest in the afternoon and
       lowest in the middle of the night.
    B) The nocturnal habits of the flying squirrel.
    C) The mating in white tailed deer.
    D) A rhythm that can be entrained.
    Answer: C

Pages: 106

12. Circadian rhythms
    A) can be entrained.
    B) are based on approximately 24 hour cycles.
    C) are free running in constant environments.
    D) all of the above
    Answer: D

Pages: 106
13. Circadian rhythms require
    A) a biological clock.
    B) a constant temperature.
    C) a constant pressure.
    D) a constant light.
    Answer: A

Pages: 106
14. The major synchronizer for the circadian rhythms is
    A) the moon.
    B) rainfall.
    C) light.
    D) temperature.
    Answer: C

Pages: 107
15. In many animals, the biological clock is found in
    A) the spinal cord.
    B) the kidney.
    C) the brain.
    D) the gonads.
    Answer: C

Pages: 107
16. The pineal gland is important as a site of the biological clock in
    A) fish and insects.
    B) bacteria and fungi.
    C) plants.
    D) birds and mammals.
    Answer: D

Pages: 110
17. Grunion, crabs, and other organisms with tidal rhythms when brought to the lab and kept under constant conditions
    A) lose their tidal rhythms.
    B) develop new tidal rhythms.
    C) keep the same tidal rhythms as they had before being removed.
    D) both A and B
    Answer: C

Pages: 108
18. A short day plant will flower only if the day length is less than the critical day length of
    A) 20 hours.
    B) 10-14 hours.
    C) 14-20 hours.
    D) 24 hours.
    Answer: B

Pages: 110
19. The study of seasonality of events is called
    A) circadian studies.
    B) circular studies.
    C) phenology.
    D) rhythm studies.
    Answer: C

True-False

Write T if the statement is true and F if the statement is false.

Pages: 100
1. Red and infrared are the only light absorbed by the ozone layer.
   Answer: False

Pages: 100
2. Leaves are green because they reflect green light and absorb red and blue light.
   Answer: True

Pages: 103
3. Sun tolerant species do not grow well in the shade.
   Answer: True

Pages: 103
4. Sun tolerant species are the first to grow in a newly cleared area.
   Answer: True

Pages: 104
5. An example of some shade tolerant species are sugar maple and beech.
   Answer: True

Pages: 102
6. Aquatic plants are mostly sun tolerant.
   Answer: False

Pages: 100
7. PAR is not used by photosynthesizing plants.
   Answer: False

Pages: 110
8. Phenological responses are due to circadian rhythms.
   Answer: False

Pages: 101
9. After being absorbed and reflected by vegetation, the wavelength of light reaching the forest floor is the same as was observed above the floor.
   Answer: False

Pages: 102
10. The compensation point is the light intensity at which photosynthesis equals respiration.
    Answer: True

Pages: 103
11. Shaded leaves of oak trees have smaller deeper lobes than sun leaves.
    Answer: False

Pages: 106
12. Among animals photoperiodism effects reproductive behavior.
    Answer: True

Pages: 100
13. PAR is visible radiation from 280nm - over 740nm.
    Answer: True

Pages: 108
14. Circadian rhythms allow an animal to predict environmental changes.
    Answer: True

Pages: 108
15. Plants that flower during the fall are short-day plants.
    Answer: True

Pages: 106
16. In many animal circadian rhythms are regulated by a clock located in the inner ears.
    Answer: False

Short Answer

Write the word or phrase that best completes each statement or answers the question.

Pages: 100
1. Photosynthetically active radiation is the _____ light of the spectrum.
   Answer: visible

Pages: 100
2. Wavelengths shorter than visible light are _____.
   Answer: ultraviolet

Pages: 100
3. Wavelengths longer than visible light are _____.
   Answer: infrared or thermal radiation

Pages: 101
4. The chloroplast absorbs _____ light.
   Answer: red and blue

Pages: 101
5. The _____ light is reflected from leaves.
   Answer: green

Pages: 101
6. In addition to spectral quality, light has three other components, intensity, duration, and

   _____.

   Answer: directionality

Pages: 103
7. Light entering a forest or prairie is _____ or modified by the vegetation.
   Answer: attenuated

Pages: 102
8. The intensity of light that will no longer allow plants to carry on sufficient photosynthesis to overcome respiratory energy is the _____.
   Answer: compensation point

Pages: 106
9. An animal's behaviors, and physiological respones in a 24 hour cycle are called _____.
   Answer: circadian rhythms

Pages: 106
10. Knowledge of circadian rhythms implies a _____.
    Answer: timekeeper or biological clock

Pages: 106
11. Circadian rhythms, once entrained to alight dark period, and continue to run even in darkness in the absence of environmental cues are called _____.
    Answer: free running

Pages: 108
12. Short-day plants will not flower if the night is interrupted by _____.
    Answer: light

Pages: 108
13. Photoperiodic phenomena are regulated by the critical _____.
    Answer: daylength

Pages: 109
14. Gonad development and spring migratory behavior increase with _____.
    Answer: daylength

Pages: 110
15. A study of the causes of timing exhibited by seasonal changes is called _____.
    Answer: phenology

Pages: 110
16. Seasonal activities in animals center around _____.
    Answer: reproduction and feeding

Matching

Choose the item from Column 2 that best matches each item in Column 1.

Pages: 110
1. Column 1: Grunions swim out to sea on high tides.
   Column 2: tidal cycle

Pages: 106
2. Column 1: Squirrels feed only during daylight and sleep at night.
   Column 2: circadian rhythm

Pages: 106

3. Column 1: Human females ovulate every 28 days.
   Column 2: lunar cycle

Pages: 110

4. Column 1: Snowshoe rabbits change from brown to white hair.
   Column 2: annual cycle

Pages: 106

5. Column 1: Light-dark experiences that set the master clock.
   Column 2: entrainment

Pages: 100

6. Column 1: Spectrum related to heat waves.
   Column 2: infrared

Pages: 100

7. Column 1: Makes up white light.
   Column 2: visible spectrum

Pages: 100

8. Column 1: Red and blue light used in photosynthesis.
   Column 2: PAR

Pages: 100

9. Column 1: Filtered out to some extent by ozone layer.
   Column 2: ultraviolet

Pages: 100

10. Column 1: Important in setting time length for animal behavior.
    Column 2: visible spectrum

Pages: 108

11. Column 1: Regulates flowering in plants.
    Column 2: photoperiod

Essay

Write your answer in the space provided or on a separate sheet of paper.

Pages: 100

1. Special lamps can be purchased to grow plants in homes during the winter. What wavelengths do these lights emit?
   Answer: Mostly red and blue (purple), the wavelengths of PAR.

Pages: 108

2. What does the term photoperiod mean and to what does it apply?
   Answer: Day length determines a phenomena like flowering in plants. There are long day, short day, and day neutral responses.

Pages: 107

3. Why are circadian rhythms important?
   Answer: They allow organisms to anticipate future events.

Pages: 103

4. What important characteristics do shade tolerant species possess that are adaptations to low light?

   Answer: Carry on photosynthesis in low light, lower rate of photosynthesis, and lower rate of respiration.

Pages: 112

5. What are the seasonal characteristics of a white tailed deer?

   Answer: Jan-Mar: antlers shed, range established,
   Apr-May: new antlers appear, June: fawning
   Jul-Sep: fawns raised, loose spots,
   Sep: Velvet off,
   Oct: necks swell,
   Nov-Dec: rutting

CHAPTER 8   Nutrients

Multiple-Choice

Choose the one alternative that best completes the statement or answers the question.

Pages: 116
1. Some soils are deficient in nutrients and addition of nutrients occurs by
    A) wetfall.
    B) wind.
    C) animals.
    D) all of the above
    Answer: D

Pages: 117
2. Nitrogen, potassium, phosphorus, and carbon are
    A) not required in ecosystems.
    B) are micronutrients.
    C) are macronutrients.
    D) are not important to consumers.
    Answer: C

Pages: 118
3. In the cycle of nutrients through plants
    A) some nutrients are lost to the soil by leaves.
    B) some nutrients move up into leaves from the soil.
    C) rain results in additional nutrients in the soil.
    D) all of the above
    Answer: D

Pages: 120
4. Which of the following is not a major source of nutrients in aquatic ecosystem.
    A) drainage water from land
    B) detritus
    C) precipitation
    D) stemflow
    Answer: D

Pages: 120
5. Iron, manganese, molybdenum, and iodine are
    A) macronutrients.
    B) micronutrients.
    C) not required by plants.
    D) not required by animals.
    Answer: B

Pages: 121
6. Calcicoles
    A) generally grow best in an acid soil.
    B) grow best in high calcium soils.
    C) are calcium deficient.
    D) are not related to calcium or pH.
    Answer: B

Pages: 122

7. Which of the following is not a calcifuge.
   A) rhododendron
   B) alfalfa
   C) azaleas
   D) all of the above
   Answer: B

Pages: 122

8. Serpentine soils
   A) are derived from ultrabasic magnesium silicate rocks.
   B) are high in magnesium and iron.
   C) are low in calcium, phosphorus, and sodium.
   D) all of the above
   Answer: D

Pages: 123

9. Halophytes
   A) grow in soils of more than 0.2% salt.
   B) are obligates to the environment of high salt.
   C) are vulnerable to high salt levels.
   D) grow only in pure water.
   Answer: A

Pages: 124

10. Grazing herbivores
   A) digest plant cellulose.
   B) prefer nitrogen-rich plants.
   C) require quality forage.
   D) all of the above
   Answer: D

Pages: 125

11. The most variable nutrient in the arctic and forest ecosystems is
   A) calcium.
   B) nitrogen.
   C) carbon.
   D) sodium.
   Answer: D

Pages: 126

12. To counteract mineral deficiencies in the spring, the large herbivores seek mineral licks for
   A) sodium.
   B) calcium.
   C) magnesium.
   D) all of the above
   Answer: D

Pages: 120
13. Nutrient cycling in an ecosystem is influenced by the organisms. As a result
    A) phytoplankton and zooplankton cycle nutrients faster than other ecosystems.
    B) phytoplankton and zooplankton cycle nutrients slower than other ecosystems.
    C) forest trees return nutrients rapidly to ecosystem.
    D) evergreens are short-term nutrient storage components in forest ecosystem.
    Answer: A

14. The important nutrient source in aquatic communities
    A) is input from surrounding land.
    B) is detritus.
    C) is precipitation.
    D) all of the above
    Answer: D

True-False

Write T if the statement is true and F if the statement is false.

Pages: 116
1. Some minerals are obtained from the soil for plant growth.
   Answer: True

Pages: 116
2. Rain water is not a source of minerals.
   Answer: False

Pages: 116
3. Carbon is not an essential element.
   Answer: False

Pages: 123
4. Halophytes grow in soil with high salt content.
   Answer: True

Pages: 125
5. The mineral content of grasses in the spring often limits growth and health of herbivores.
   Answer: False

Pages: 124
6. Sulphur dioxide in the atmosphere can destroy vegetation.
   Answer: True

Pages: 123
7. All halophytes are obligate organisms.
   Answer: False

Pages: 125
8. Sodium dificiencies have been found in Australian herbivores, African elephants, white tailed deer, and moose.
   Answer: True

9. Herbivores show some preference for most nitrogen rich plants.
   Answer: True

10. When minerals are scarce herbivores stop eating.
    Answer: False

11. To counteract mineral imbalance in the spring large herbivores seek out mineral licks.
    Answer: True

12. Endemics are species which are restricted to specialized habitat.
    Answer: True

13. There have been no evolutionary changes to produce species tolerant to toxic heavy metals like copper.
    Answer: False

14. Throughfall and stemflow do not supply many useful elements for tree growth.
    Answer: False

15. Nutrients such as nitrogen, phosphorus, potassium, and calcium frequently cycle through trees to the ground and back into trees.
    Answer: True

16. The major sources of nutrients for aquatic life are drainage water, detritus, sediment, and precipitation.
    Answer: True

Short Answer

Write the word or phrase that best completes each statement or answers the question.

1. Nitrogen, carbon, and phosphorus are _____ because they are required in large amounts.
   Answer: macronutrients

2. Elements required in trace amounts such as copper, zinc, and boron are said to be _____.
   Answer: micronutrients

3. All consumer organisms from carnivores to decomposers depend on _____.
   Answer: green plants

Pages: 116
4. Many nutrients in soil are carried there by rain or snow; these are called _____.
   Answer: wetflow

Pages: 116
5. Mineral ions reach the soil and roots by _____ and _____.
   Answer: stemflow and throughfall   (any order)

Pages: 119
6. Considerable nutrients are withdrawn permanently from ecosystems by harvesting and these must be replaced by _____.
   Answer: fertilizing

Pages: 120
7. Detritus, small particles of organic matter, are an important source of nutrients in _____ systems.
   Answer: aquatic

Pages: 120
8. Plants require _____ essential elements.
   Answer: 16

Pages: 121
9. In grassland areas addition of nitrogen discourages growth of _____.
   Answer: legumes

Pages: 121
10. Plants requiring high lime in the soil to grow are called _____.
    Answer: calcicoles

Pages: 122
11. _____ soils are derived from ultrabasic magnesium silicate rocks which are green in color and contain much iron but are low in calcium.
    Answer: Serpentine

Pages: 122
12. Species restricted to specialized habitats are _____.
    Answer: endemics

Pages: 123
13. Soils high in sodium chloride whether on deserts or along oceans are called _____.
    Answer: saline

Pages: 125
14. Mineral deficiencies are common in plants and animals in the _____ of the year.
    Answer: spring

Pages: 127
15. Reproductive success in deer is related to mineral levels on good _____.
    Answer: range land

Pages: 121
16. Inflow and outflow of nutrients make up the community _____.
    Answer: nutrient budget

Matching

Choose the item from Column 2 that best matches each item in Column 1.

Pages: 117
1. Column 1: In plants, it is necessary for sugar and starch formation. In animals, it is involved in nerve and muscle function, protein formation, and growth.
   Column 2: potassium

Pages: 117
2. Column 1: In plants and animals, it is a major constituent of protein.
   Column 2: nitrogen

Pages: 117
3. Column 1: A basic constituent of living matter; recipient of hydrogen in aerobic respiration.
   Column 2: oxygen

Pages: 117
4. Column 1: The basic constituent of all living matter.
   Column 2: carbon

Pages: 117
5. Column 1: In animals it is used for clotting of blood, acid-base balance, contraction and relaxation of heart muscles.
   Column 2: calcium

Essay

Write your answer in the space provided or on a separate sheet of paper.

Pages: 116
1. What macronutrients are important and why are they important?
   Answer: nitrogen, calcium, carbon, oxygen, phosphorus, potassium, and sodium. All these nutrients are essential for growth and function of all life.

Pages: 116
2. What are some important micronutrients?
   Answer: copper, zinc, iron, magnesium, and sulfur

Pages: 127
3. What is a nutrient budget? Give an example.
   Answer: Nutrients are added, stored, and lost from any ecosystem, representing inputs and outputs. A measure of this flow is a nutrient budget. Any example selected from text will do. Nutrient sources are soil, wet and dry atmospheric depositions. Nutrients are recycled through the ecosystem by deposition leaching, uptake by plants, consumption, decomposition, short and long term storage, dead organic matter and soil.

Pages: 124
4. What effect does the nutrient level of plants have on the well-being of animals?
   Answer: Plants are the major nutrient source of all animals, directly or indirectly. Herbivores are particulary sensitive to nutrient levels in plants, especially proteinous nitrogen, essential for growth and reproduction.

CHAPTER 9   Soils

Multiple-Choice

Choose the one alternative that best completes the statement or answers the question.

Pages: 138
1. Soils developed on wind blown parent materials are called
   A) till soils.
   B) bedrock soils.
   C) alluvial soils.
   D) loess soils.
   Answer: D

Pages: 131
2. Consolidated bedrock from which soil is developed is called the
   A) till.
   B) R horizon.
   C) A horizon.
   D) all of the above
   Answer: B

Pages: 131
3. Whitish and greyish colors in soils come from
   A) quartz.
   B) kaolin.
   C) carbonates of calcium and magnesium.
   D) all of the above
   Answer: D

Pages: 132
4. Soil texture is based on _____ content.
   A) sand
   B) silt
   C) clay
   D) all of the above
   Answer: D

Pages: 130
5. Soils are found in layers called
   A) parent materials.
   B) topographic positions.
   C) horizons.
   D) all of the above
   Answer: C

Pages: 135
6. Because of its unique properties and structure, _____ controls many important properties of soil such as cation exchange capacity.
   A) clay
   B) sand
   C) silt
   D) gravel
   Answer: A

Pages: 137
7. Bacteria, protozoans, fungi, and _____ are found in soil.
   A) insects
   B) mites
   C) nematodes
   D) all of the above
   Answer: D

Pages: 139
8. The partially decomposed organic matter in soil is called
   A) parent material.
   B) clay.
   C) humus.
   D) all of the above
   Answer: C

Pages: 138
9. The breakdown of materials in the <u>R</u> horizon into soil is called
   A) humus formation.
   B) weathering.
   C) dribbling.
   D) leaching.
   Answer: B

Pages: 130
10. No litter or humus is found in the
    A) <u>O</u> horizon.
    B) <u>A</u> horizon.
    C) <u>B</u> horizon.
    D) <u>R</u> horizon.
    Answer: D

Pages: 130
11. The _____ is the zone with the most biological activity.
    A) <u>O</u> horizon
    B) <u>A</u> horizon
    C) <u>B</u> horizon
    D) <u>C</u> horizon
    Answer: B

Pages: 131
12. The maximum accumulation of clay minerals occurs in the
   A) O̲ horizon.
   B) A̲ horizon.
   C) B̲ horizon.
   D) C̲ horizon.
   Answer: C

Pages: 130
13. In soils with sufficient rainfall leaching occurs when
   A) twigs are decomposed by fungi.
   B) humus is decomposed by bacteria.
   C) soluble nutrients are carried away by water.
   D) soils are waterlogged and do not drain.
   Answer: C

Pages: 138
14. Parent material from bedrock is formed from
   A) igneous rock.
   B) sedimentary rock.
   C) metamorphic rock.
   D) all of the above
   Answer: D

Pages: 139
15. Mull soils
   A) have all organic material converted to true humic substances.
   B) have a dense layer of litter.
   C) are prairie soils.
   D) have no A̲ horizon.
   Answer: A

Pages: 139
16. Agricultural crops might be expected to grow best on
   A) saline soils.
   B) vertisols.
   C) clay pens.
   D) mollisols.
   Answer: D

Pages: 143
17. The United States system of classifying soils uses _____ main orders.
   A) 100
   B) 20
   C) 11
   D) 7
   Answer: C

Pages: 144
18. Erosion by water is increased by
    A) heavier rainfall.
    B) compaction.
    C) removal of vegetation.
    D) all of the above
    Answer: D

Pages: 130
19. The R horizon is composed of
    A) humus.
    B) soil.
    C) material from the B horizon.
    D) bedrock.
    Answer: D

Pages: 132
20. The arrangement of aggregates or peds is called
    A) soil texture.
    B) soil series.
    C) soil structure.
    D) soil series.
    Answer: C

True-False

Write T if the statement is true and F if the statement is false.

Pages: 130
1. Soil is the mineral material between the vegetation and the parent rock.
   Answer: True

Pages: 138
2. At the entrance of rivers into larger lakes or rivers, large deposits of sediments occur called an alluvial deposit.
   Answer: True

Pages: 131
3. The rocky material upon which soil is developed is called the B horizon.
   Answer: False

Pages: 130
4. In forest soils the A horizon is usually alkaline.
   Answer: False

Pages: 142
5. Soils of the prairies with calcification and a thick horizon developed because of heavy rainfall.
   Answer: False

Pages: 139

6. Since mull humus is nearly alkaline bacteria, annelids, termites, and Collembola are involved in the breakdown of the humus.
   Answer: True

Pages: 140

7. Clay is composed of aluminum silicate.
   Answer: True

Pages: 130

8. In the soil profile, the <u>O</u> is the horizon of accumulation of minerals.
   Answer: False

Pages: 135

9. Exchangeable cations are attached to sand particles.
   Answer: False

Pages: 148

10. Road building, construction, and agriculture help prevent erosion.
    Answer: False

Pages: 140

11. Red soils in good rainfall areas of the Southeastern U.S. are called laterites.
    Answer: True

Pages: 132

12. Soil aggregates are called peds.
    Answer: True

Pages: 143

13. Soils which are poorly drained in cold climates are usually black in color.
    Answer: True

Pages: 140

14. Podzolization refers to the accumulation of calcium ions in upper layers of soil.
    Answer: False

Pages: 130

15. Soil is the foundation of terrestrial communities.
    Answer: True

Pages: 139

16. Due to the rainfall of the forested regions, organic matter is completely broken down each year so that no litter or humus accumulates.
    Answer: False

Pages: 142

17. Salinization refers to the leaching of salt from the soil.
    Answer: False

Pages: 137

18. Collembolae (springtails) are the most generally distributed of all insects in the soil.
    Answer: True

19. Moder humus is midway between mor and mull humus.
Answer: True

20. Podzols are a prairie soil type.
Answer: False

21. Despite variations in slope, climate, and native vegetation on the same parent material, the soil which develops is the same in all cases because the parent material is the same.
Answer: False

Short Answer

Write the word or phrase that best completes each statement or answers the question.

1. The stratum below the vegetation and above the hard rock is called _____.
Answer: soil

2. A natural product from weathered rock is called _____.
Answer: soil

3. Layers in the soil which differ in appearance, chemical reactions, and physical characteristics are called _____.
Answer: horizons

4. The soil _____ is the appearance of a cross section of soil.
Answer: profile

5. The R horizon is composed of the _____ upon which the soil is developed.
Answer: bedrock

6. The horizon with remains of plants and animals some of which are decomposed is the _____.
Answer: O horizon

7. The _____ color in soil is due to ferric oxide.
Answer: red

8. _____ relates to pore space in soil.
Answer: Texture

9. Texture is determined by the sand, _____, and _____ content of the soil.
Answer: silt, clay   (any order)

10. The arrangement of aggregates or peds is called _____.
Answer: soil structure

11. The water held in the soil between field capacity and the permanent wilting percent is the _____ for plants.
Answer: available water capacity

12. Clay forms micelles which attract certain _____ making the cation exchange capacity of the soil.
Answer: positively charged ions

13. Protoqoans, bacteria, fungi, mites, Collembola, and earthworms are the agents of _____ forming _____ or organic matter.
Answer: decay, humus   (in order)

14. _____, _____, and _____ rocks form the parent material for soil and determine some of the characteristics of the final material.
Answer: igneous, sedimentary, metamorphic   (any order)

15. _____ is a soil of deciduous forests with calcium and with distinct horizons.
Answer: Mull

16. Heavy rainfall and warm tropical climates produce a red _____ soil.
Answer: laterite

17. Soils common to arid and semi-arid environments have developed through the process of _____.
Answer: salinization

18. Soils with similar horizons except the A horizon belong to a _____.
Answer: soil series

19. The movements of soil particles by wind or water is called _____.
Answer: soil erosion

20. Soils which are low in calcium and high in heavy metals such as nickel and magnesium are _____ soils.
Answer: serpentine

21. _____ takes place under poor drainage and produces black humus soils under tundra-like conditions.
Answer: Gleyization

Matching

Choose the item from Column 2 that best matches each item in Column 1.

Match the following horizons with their descriptions.

Pages: 130
1. Column 1: Little or no organic matter or parent material found in this layer.
   Column 2: <u>C</u> horizon

Pages: 131
2. Column 1: Accumulation of minerals in this layer.
   Column 2: <u>B</u> horizon

Pages: 131
3. Column 1: Parent materials in this layer.
   Column 2: <u>R</u> horizon

Pages: 130
4. Column 1: Organic debris not broken down is found in this layer.
   Column 2: <u>O</u> horizon

Pages: 130
5. Column 1: This layer is composed of topsoil.
   Column 2: <u>A</u> horizon

Match the following.

Pages: 139
6. Column 1: Humus with nearly neutral pH and broken down by bacteria, insects, and mites.
   Column 2: mull

Pages: 140
7. Column 1: Development of red soil from high rainfall and warm climates where there is a high ferric oxide content in the soil.
   Column 2: laterization

Pages: 139
8. Column 1: Humus of acid nature breaks down mostly by fungi.
   Column 2: mor

Pages: 140
9. Column 1: Process in soil development where depletion of bases and movement of iron and clays occurs downward.
   Column 2: podzolization

Pages: 142
10. Column 1: Process of soil development in arid to plains areas where calcium accumulates.
    Column 2: calcification

Essay

Write your answer in the space provided or on a separate sheet of paper.

Pages: 147
1. What factors led to the dust bowl of the 1930's?
   Answer: Drought, removal of natural vegetation for crops, overgrazing, plowing every year, and breaking up the <u>A</u> horizon.

Pages: 130
2. What is soil?
   Answer: Weathered mineral matter and organic materials capable of supporting plant growth. It is formed from weathered rock materials by action of climate and living organisms.

Pages: 132
3. What determines the texture of soil?
   Answer: Texture is determined by the sand, silt, and clay proportions in soil and the humus. Loam soil is equal parts sand, silt, and clay. There are sandy soils, silty soils, and clay soils as well as silty clay loam and others.

Pages: 130
4. Why is soil the medium on which terrestrial plants grow?
   Answer: 1. Holds moisture
   2. Provides mechanical support
   3. Contains essential elements

CHAPTER 10 Concept of the Ecosystem

Multiple-Choice

Choose the one alternative that best completes the statement or answers the question.

Pages: 154
1. The light reaction of photosynthesis involves
    A) fixation of carbon dioxide into sugars.
    B) conversion of light to chemical energy of NADP and ATP.
    C) utilization of oxygen for carbon dioxide production.
    D) ribulose bisphosphate and carbon dioxide reaction.
    Answer: B

Pages: 154
2. The C3 pathway of photosynthesis
    A) occurs in most green plants and algae.
    B) does not fix carbon dioxide.
    C) uses oxygen to incorporate organic acids.
    D) is found in all grasses.
    Answer: A

Pages: 154
3. At low carbon dioxide levels, photorespiration occurs in which _____ and not carbon dioxide is attached to ribulose bisphosphate.
    A) hydrogen
    B) carbon
    C) nitrogen
    D) oxygen
    Answer: D

Pages: 154
4. Plants with C4 can carry on photosynthesis more efficiently in
    A) cool climates.
    B) high temperature.
    C) high carbon dioxide concentration.
    D) low light intensity.
    Answer: B

Pages: 155
5. Succulent green plants usually found in the desert usually use
    A) C3 photosynthesis.
    B) C4 photosynthesis.
    C) CAM photosynthesis.
    D) photorespiration and not photosynthesis.
    Answer: C

Pages: 156
6. C4 plants are usually found where _____ is limited such as grasslands.
   A) water
   B) light
   C) nitrogen
   D) oxygen
   Answer: A

Pages: 166
7. Photosynthetic organisms are the _____ in an ecosystem.
   A) consumers
   B) heterotrophs
   C) decomposers
   D) autotrophs
   Answer: D

Pages: 169
8. Which of the following is not a heterotroph?
   A) mushroom
   B) earthworm
   C) moose
   D) algae
   Answer: D

Pages: 166
9. An abiotic component of the ecosystem is
   A) green plants.
   B) omnivores.
   C) light
   D) decomposers.
   Answer: C

Pages: 166
10. Which of the following is not an autotroph?
    A) moose
    B) maple tree
    C) prairie grass
    D) shrubs
    Answer: A

Pages: 152
11. The hydrosphere, lithosphere, biosphere, and the atmoshpere are collectively called the
    A) troposphere.
    B) mesosphere.
    C) ecosphere.
    D) shereosphere.
    Answer: C

Pages: 161
12. Leaching, fragmentation, and catabolism are stages of
    A) leaf build up.
    B) soil formation from bedrock.
    C) carnivore increases.
    D) decomposition.
    Answer: D

Pages: 164
13. In aquatic systems
    A) humus is an important intermediate in decomposition.
    B) detritus is important in decomposition.
    C) organic matter does not decompose.
    D) fecal material is not involved.
    Answer: B

Pages: 161
14. On the ground litter is attacked by
    A) fungi.
    B) bacteria.
    C) insects.
    D) all of the above
    Answer: D

Pages: 152
15. The concept of an ecosystem was first coined by
    A) R.L. Smith.
    B) C. Darwin.
    C) J. Von Liebig.
    D) A.E. Tansley
    Answer: D

Pages: 159
16. Decomposition is influenced by
    A) abiotic factors.
    B) herbivores.
    C) carnivores.
    D) all of the above
    Answer: D

True-False

Write T if the statement is true and F if the statement is false.

Pages: 152
1. The ecosystem is composed of producers, consumers, and abiotic factors.
   Answer: True

Pages: 153
2. Heterotrophs are green plants.
   Answer: False

Pages: 153
3. Autotrophs include such animals as insects.
   Answer: False

Pages: 153
4. Autotrophs carry on photosynthesis.
   Answer: True

Pages: 153
5. Green plants require carbon dioxide and water to synthesize sugar.
   Answer: True

Pages: 154
6. The two stages of photosynthesis are the light reactions and the dark reactions.
   Answer: True

Pages: 154
7. The dark reaction is the formation of ATP and NADP by the chloroplast.
   Answer: False

Pages: 154
8. Carbon dioxide is converted to sugars in the dark phase of photosynthesis.
   Answer: True

Pages: 155
9. The C4 system does not operate under low carbon dioxide concentrations when the stomata are closed.
   Answer: False

Pages: 155
10. Plants using the C4 system have an advantage over plants using the C3 system under hot dry conditions.
    Answer: True

Pages: 154
11. Forest plants and green algae usually use the C3 system.
    Answer: True

Pages: 156
12. CAM plants open stomata in the daytime and close the stomata at night for carbon dioxide uptake.
    Answer: False

Pages: 156
13. C3 plants may be more competitive than C4 plants.
    Answer: False

Pages: 164
14. Some autotrophs are insects, fungi, and bacteria.
    Answer: False

Pages: 156
15. CAM plants are usually succulents and cacti.
    Answer: True

Short Answer

Write the word or phrase that best completes each statement or answers the question.

Pages: 152
1. The thin blanket of life surrounding the earth is the _____.
   Answer: biosphere

Pages: 152
2. The body of liquid on the earth is the _____.
   Answer: hydrosphere

Pages: 152
3. The nutrient-regenerating and energy-processing system is called the _____.
   Answer: ecosystem

Pages: 153
4. The biotic part of the system composed of interacting organisms is the _____.
   Answer: community

Pages: 153
5. The energy trapping base of large green plants and algae make up the _____ in the ecosystem.
   Answer: producers

Pages: 153
6. The heterotrophs which utilize the food produced by the green plants and algae are composed of two classes, the _____ and the _____.
   Answer: consumers, decomposers   (any order)

Pages: 153
7. The third component consists of _____, dissolved organic and inorganic materials in aquatic systems, and the soil matrix.
   Answer: dead organic matter

Pages: 154
8. Plants using the Calvin-Benson pathway of carbon fixation by using phosphoglycolate to fix carbon dioxide are called _____.
   Answer: C3 plants

Pages: 155
9. Plants from warmer drier regions use the _____ and the _____.
   Answer: C4 pathway, CAM pathway   (any order)

Pages: 155
10. The LAI refers to the _____ which states that the lower this value is, the more light reaches the ground.
    Answer: leaf area index

Pages: 159
11. Fragmented leaves, twigs and other small pieces of organic matter are called _____.
    Answer: detritus

Pages: 160
12. The decomposers are _____, _____, and _____.
    Answer: microflora, detritavores, micro herbivores   (any order)

Pages: 161
13. The four stages of decomposition are _____ (name all four).
    Answer: leaching, fragmentation, catabolism, mineralization
             (any order)

Matching

Choose the item from Column 2 that best matches each item in Column 1.

Pages: 153
1. Column 1: Herbivores
   Column 2: consumers

Pages: 153
2. Column 1: Photosynthetic organisms
   Column 2: producers

Pages: 153
3. Column 1: Autotrophs
   Column 2: producers

Pages: 153
4. Column 1: Fungi
   Column 2: consumers

Pages: 153
5. Column 1: Carnivores
   Column 2: consumers

Pages: 153
6. Column 1: Sunlight
   Column 2: abiotic

Essay

Write your answer in the space provided or on a separate sheet of paper.

Pages: 154
1. How do the light and dark reactions work in photosysthesis?
   Answer: Light energy is trapped by the chloroplasts and the energy is used to form NADPH and
            ATP. The energy of ATP is used to drive the addition of hydrogen atoms and formation
            of sugars from carbon dioxide.

Pages: 154

2. How does C3 photosynthesis differ from C4 photosynthesis?

Answer: In C3 photosynthesis, carbon dioxide reacts with ribulose bisphosphate and the product is converted to glucose. In C4, the carbon dioxide reacts with organic acids to form 4 carbon acids and these loose the carbon dioxide to the C3 cycle. Thus, there is an extra step in the C4 cycle.

Pages: 153

3. What are the three components of an ecosystem?

Answer: The components are producers (autotrophs), consumers (herbivores, carnivores, and decomposers), and the abiotic factors (light, weather, minerals).

Pages: 159

4. Why are the decomposers important and what are the major groups of decomposers?

Answer: Decomposition is important to prevent the buildup of organic matter. Microorganisms, insects, arthropods, and earthworms are the most important terrestrial organisms and bacteria, protoqoans and insects for aquatic systems.

Pages: 165

5. What is POM and why is it important?

Answer: POM is particulate organic matter and it is important in microbial decomposition in aquatic habitats and as a food source for crustaceans.

CHAPTER 11 Ecosystem Energetics

Multiple-Choice

Choose the one alternative that best completes the statement or answers the question.

Pages: 169
1. Organisms of gross primary production are
   A) autotrophs.
   B) consumers.
   C) decomposers.
   D) saprophagous insects.
   Answer: A

Pages: 168
2. The ultimate energy for living systems is
   A) starch molecules.
   B) detritus.
   C) proteins.
   D) sunlight.
   Answer: D

Pages: 175
3. The highest net primary productivity occurs in
   A) tropical rain forests.
   B) the middle of the ocean.
   C) boreal forests.
   D) tundra.
   Answer: A

Pages: 177
4. Primary productivity can be estimated by
   A) light and dark bottle.
   B) radio carbon tracer.
   C) chlorophyll concentration.
   D) all of the above
   Answer: D

Pages: 181
5. The energy budget of a homeotherm consumer has an assimilation efficiency close to
   A) 23%.
   B) 34%.
   C) 77%.
   D) 100%.
   Answer: C

Pages: 191
6. A food pyramid consists of several feeding or trophic levels. The first one is
   A) decomposers.
   B) producers.
   C) herbivores.
   D) carnivores.
   Answer: B

Pages: 191

7. The last trophic level in a food pyramid are
   A) producers.
   B) primary consumers.
   C) top carnivores.
   D) herbivores.
   Answer: C

Pages: 191

8. If a small patch of wheat had 100 calories, how much would go to the herbivores.
   A) 100
   B) 50
   C) 10
   D) 1
   Answer: C

Pages: 191

9. In a forest community, which trophic level has the least biomass.
   A) producer
   B) primary consumer
   C) herbivore
   D) top carnivore
   Answer: D

Pages: 169

10. There is a decrease in energy through a food chain because
    A) the first law thermodynamics.
    B) the second law of thermondynamics.
    C) it is not dependent on thermodynamics.
    D) energy does not behave as stated in a food chain.
    Answer: B

Pages: 191

11. A bulk of the energy in an ecosystem is found in the
    A) primary producers.
    B) secondary consumers.
    C) herbivores.
    D) decomposers.
    Answer: A

Pages: 185

12. In a yellow poplar forest, 50% of the gross primary productivity goes into maintenance and
    respiration. How much goes to the detritus food chain?
    A) 50%
    B) 35%
    C) 25%
    D) 10%
    Answer: B

13. Detrital food chains have area of accumulation called
    A) maintenance areas.
    B) sinks.
    C) small arachnid stomachs.
    D) food webs.
    Answer: B

14. In terrestrial systems, only a small part of the energy goes by the way of
    A) saprophagous food chains.
    B) detrital food chains.
    C) grazing food chains.
    D) major food chains.
    Answer: C

15. Energy consumed and assimilated goes to
    A) maintenance.
    B) growth.
    C) reproduction.
    D) all of the above
    Answer: D

True-False

Write T if the statement is true and F if the statement is false.

1. The watt is a measure of energy.
   Answer: False

2. Net primary production is the amount of energy left after respiration.
   Answer: True

3. Net primary production is not used for growth, storage or other uses; it is for the use of primary consumers only.
   Answer: False

4. Biomass is the amount of energy fixed by primary production.
   Answer: False

5. Heat can be used to do work by cells.
   Answer: False

6. The energy made available to organisms is either used to do work or make new molecules.
   Answer: True

7. Aquatic communities are most efficient at the compensation point.
   Answer: False

8. One of the most useful and sensitive methods of determining productivity is the radiocarbon tracer method.
   Answer: True

9. One of the most useful methods in aquatic systems is the carbon flux method of determining productivity.
   Answer: False

10. Secondary productivity is equal to the amount consumed minus respiration energy minus energy in urine energy and in feces.
    Answer: True

11. Decomposers are the final consumer group.
    Answer: True

12. Within any community there are two major food chains, the grazing food chain, and the detrital food chain.
    Answer: True

Short Answer

Write the word or phrase that best completes each statement or answers the question.

1. _____ is stored energy at rest which can do work if the situation changes.
   Answer: Potential energy

2. The _____ law of thermodynamics states that energy can be converted but it can never be created nor destroyed.
   Answer: first

3. Photosynthesis is an _____ reaction.
   Answer: exothermic

4. The total amount of photosynthesis by green plants in energy terms is the _____.
   Answer: gross primary production

5. _____ is gross primary production minus green plant respiration and represents the energy available for heterotrophe.
   Answer: Net primary production

6. The accumulated organic matter found in an area at some time is the _____.
   Answer: standing crop biomass

7. Of net primary production, some are allocated for _____, _____, and _____.
   Answer: growth, storage, accumulation   (any order)

8. Temperature and rainfall influence _____.
   Answer: primary productivity

9. A _____ is a series of movements of energy from one feeding group to another as one group eats another.
   Answer: food chain

10. Organisms that eat animals are _____ and plant eating organisms are _____.
    Answer: carnivores, herbivores   (in order)

11. By representing food chains as a pyramid of _____ levels, a picture of energy loss per layer can be observed.
    Answer: trophic

12. A carnivore feeding on a herbivore is a/an _____.
    Answer: second level consumer

13. _____ is the dead organic matter attacked by saprophagous insects.
    Answer: Detritus

14. No transfer of energy is 100% efficient; some heat and entropy are lost in spontaneous reactions. This is a statement of _____.
    Answer: the second law of thermodynamics

15. The transfer of energy from one trophic level to another is limited to about _____; the rest is lost as heat or passed as feces.
    Answer: 10%

16. Food chains interacting form _____; a better description of the community.
    Answer: food webs

Matching

Choose the item from Column 2 that best matches each item in Column 1.

Pages: 185
1. Column 1: Litter on the forest floor.
   Column 2: detritus

Pages: 169
2. Column 1: Dominant trees
   Column 2: producers

Pages: 169
3. Column 1: Sun flecks
   Column 2: abiotic

Pages: 181
4. Column 1: Aphids on the leaves.
   Column 2: primary consumers

Pages: 182
5. Column 1: Owl on a limb.
   Column 2: secondary consumers

Pages: 184
6. Column 1: Fungi and bacteria
   Column 2: decomposers

Essay

Write your answer in the space provided or on a separate sheet of paper.

Pages: 169
1. How do the first and second law of thermodynamics determine energy movement in ecosystems communities?
   Answer: The first law states that energy is neither created or destroyed in a reaction. All the energy involved can be accounted for.

   The second law states that in any energy transfer, a fraction of the energy I lost as waste. It cannot be transferred or passed on any further. Ecologically, the law explains why less energy is available to each trophic level. Much of the energy passed from one trophic level to another is lost as heat of metabolism.

Pages: 169
2. What is the difference between net primary production and gross primary production?
   Answer: Production refers to the energy accumulated by organisms. Therefore, the gross primary production refers to all of the sun's energy which is assimilated and the net primary production is this value minus respiration.

Pages: 179

3. What is assimilation efficiency?

   Answer: The assimilation efficiency in animals is the equivalent to gross productivity in green plants. It is the percentage of food energy consumed that is converted to animal tissue.

Pages: 187

4. Why is the detrital food chain important?

   Answer: In many cases, more energy goes through the food web by the detrital food web than by the grazing food web, especially in forests and grasslands.

Pages: 192

5. What are biomass pyramids and energy pyramids and what do they tell us?

   Answer: Ecological pyramids represent the relationship of one trophic level to another. The pyramid of biomass depicts the weight of biomass as each feeding level. The pyramid of energy depicts the quantity of energy fixed, stored and passed on to the next trophic level. It emphasizes the role organisms play in the transfer of energy.

CHAPTER 12 Biogeochemical Cycles

Multiple-Choice

Choose the one alternative that best completes the statement or answers the question.

Pages: 196
1. Essential elements are stored in the earth and the atmosphere in areas called
   A) communities.
   B) storage tanks.
   C) holes.
   D) reservoirs.
   Answer: D

Pages: 197
2. The major reservoir of oxygen is not one of the following.
   A) molecular oxygen
   B) carbon dioxide
   C) ozone
   D) water
   Answer: C

Pages: 200
3. A major reservoir of carbon dioxide is
   A) the ocean.
   B) organic compounds.
   C) living organisms.
   D) decaying detritus.
   Answer: A

Pages: 202
4. Globally, carbon dioxide varies seasonally in the following reservoir.
   A) oil and gas
   B) limestone
   C) atmosphere
   D) humus
   Answer: C

Pages: 210
5. Cycles with two abiotic phases, a salt solution phase and a rock phase,
   A) are called sedimentary cycles.
   B) include sulfur.
   C) include phosphorus.
   D) all of the above
   Answer: D

Pages: 206

6. The reservoir of nitrogen is the atmosphere where it composes _____% of the gases in the atmosphere.
   A) 31
   B) 49
   C) 79
   D) 99
   Answer: C

Pages: 206

7. Nitrogen in its gaseous state is not usable by plants so it must be converted. Part of the nitrogen cycle is devoted to this. The part which is NOT involved in making nitrogen available to plants is
   A) biological fixation.
   B) nitrification.
   C) denitrification.
   D) none of the above
   Answer: C

Pages: 207

8. _____ are found in legumes and can incorporate nitrogen gas directly.
   A) Algae
   B) Bacteria
   C) Decomposers
   D) Fungi
   Answer: B

Pages: 207

9. Which of the following organisms can fix nitrogen from the atmosphere?
   A) azotabacter
   B) clostridium
   C) nostoc
   D) all of the above
   Answer: D

Pages: 213

10. Increased _____ added to the streams by agriculture brings about an excess of algal growth.
    A) carbon dioxide
    B) phosphorus
    C) oxygen
    D) nitrogen
    Answer: B

Pages: 207

11. The process in which nitrates are converted to nitrogen gas and released to the atmosphere is
    A) fixation.
    B) ammonification.
    C) mineralization.
    D) denitrification.
    Answer: D

Pages: 213
12. Heavy metal toxins could include
    A) lead.
    B) mercury.
    C) cadmium.
    D) all of the above
    Answer: D

Pages: 214
13. Destructive acidic materials in the atmosphere include
    A) nitrogen gas.
    B) carbon.
    C) nitrous oxide and sulfur dioxide.
    D) ozone.
    Answer: C

Pages: 218
14. DDT and PCBs are destructive because it is
    A) soluble in fats.
    B) soluble in oils.
    C) soluble in lipids.
    D) all of the above
    Answer: D

15. The author of <u>Silent Spring</u>, who sounded the warning in no uncertain terms about DDT and insecticides, was
    A) C. Cottam.
    B) R.L. Smith.
    C) L. Von Liebig.
    D) R. Carson.
    Answer: D

True-False

Write T if the statement is true and F if the statement is false.

Pages: 199
1. Green plants produce carbon dioxide and utilize oxygen when they respire.
   Answer: True

Pages: 204
2. A principal reservoir for carbon dioxide is the ocean.
   Answer: True

Pages: 197
3. The ozone layer is the main reservoir for oxygen in the oxygen cycle.
   Answer: False

Pages: 200
4. A major source of carbon dioxide is from the burning of fossil fuels.
   Answer: True

5. Oxygen, although necessary for life, is also toxic for anaerobic bacteria.
   Answer: True

6. The carbon dioxide concentration is nearly constant in the atmosphere.
   Answer: False

7. Nitrogen fixation converts nitrogen gas to ammonia and nitrates.
   Answer: True

8. Nitrogen content in the atmosphere is about 30%.
   Answer: False

9. The blue green algae Nostoc will fix nitrogen.
   Answer: True

10. The sulfur cycle has two components, gaseous with sulfur dioxide and sedimentary.
    Answer: True

11. The main reservoir of the phosphorus cycle is the atmosphere.
    Answer: False

12. Addition of phosphorus to lakes causes an algal bloom and eutrophication.
    Answer: True

13. Heavy metals such as lead, mercury, and cadmium are beneficial to living systems.
    Answer: False

14. Acid rain destroys plants immediately on contact but over long periods it is beneficial.
    Answer: False

15. DDT is illegal in the U.S.; therefore, it is not a significant problem.
    Answer: False

Short Answer

Write the word or phrase that best completes each statement or answers the question.

1. Mineral cycling cannot happen without _____.
   Answer: energy

Pages: 196

2. The movement of minerals through the ecosystem is called _____.
   Answer: biogeochemical cycling

Pages: 197

3. The oxygen, carbon dioxide, nitrogen, and hydrogen cycles are _____ which have their main reservoir in the _____.
   Answer: gaseous cycles, atmosphere   (in order)

Pages: 196

4. Nutrient cycling occurs with _____.
   Answer: energy flow

Pages: 196

5. Oxygen is released into the atmosphere from _____.
   Answer: photosynthesis

Pages: 197,198

6. Oxygen is used in _____, oxidation of rocks, and decomposition and it is important in the _____ layer in the atmosphere.
   Answer: respiration, ozone   (in order)

Pages: 199

7. The carbon found in living organisms and fossils is from _____.
   Answer: carbon dioxide

Pages: 205

8. Carbon dioxide in the atmosphere causes warming of the earth due to the _____ effect.
   Answer: greenhouse

Pages: 206

9. The most conspicuous nitrogen-fixing plants are _____ in which nitrogen is converted to _____.
   Answer: legumes, ammonia   (in order)

Pages: 207

10. Ammonia is converted to _____ by bacteria before use by plants.
    Answer: nitrates

Pages: 211

11. The sulfur cycle has two gaseous components: _____ and _____.
    Answer: hydrogen sulfide, sulfur dioxide   (any order)

Pages: 213

12. Phosphorus enrichment like nitrate enrichment causes _____ of aquatic systems.
    Answer: eutrophication

Pages: 213

13. Human activities have increased the cycling of lead, mercury, and cadmium that are called _____ and which are very toxic.
    Answer: heavy metals

Pages: 216
14. Acid depositions as _____ injure forest trees.
    Answer: acid rain

Pages: 216
15. _____ is an insecticide which is illegal to use because of its high toxicity to animal life, including humans and its ability to remain in the environment and pass through food chains.
    Answer: DDT

Matching

Choose the item from Column 2 that best matches each item in Column 1.

Pages: 217
1. Column 1: Toxin which does not cycle
   Column 2: DDT

Pages: 197
2. Column 1: Cycle with a gas used in respiration
   Column 2: oxygen cycle

Pages: 213
3. Column 1: Heavy metal toxin
   Column 2: mercury

Pages: 214
4. Column 1: Cycle with an oxide in the atmosphere from auto pollution
   Column 2: nitrogen cycle

Pages: 210
5. Column 1: A sedimentary and gaseous component in this cycle
   Column 2: sulfur cycle

Essay

Write your answer in the space provided or on a separate sheet of paper.

Pages: 207
1. In the nitrogen cycle, the gaseous nitrogen must be converted to nitrate for plants to use. How does this occur?
    Answer: Gaseous nitrogen is first converted to usable form, amino acids (nitrification). Then formation of ammonia (ammonification) and finally nitrification, the conversion of ammonia to nitrates. Cyanobacteria, Clostridium, and Rhizobium among other bacteria fix nitrogen.

Pages: 217
2. Why are DDT and PCBs so toxic?
    Answer: They dissolve and accumulate in fatty parts of living organisms. They are distributed in oil where they form surface layers in oceans and aquatic systems and move locally and globally. They are not broken down in the food chain and accumulate in top carnivores in high concentrations. As a consequence their toxicity is enhanced.

Pages: 200

3. What are the daily and seasonal patterns of the atmosphere? Is the atmosphere a large reservoir for carbon dioxide?

Answer: Carbon dioxide concentrations in the atmosphere surrounding green plants drops in the afternoon and go up in the evening and at night. Seasonal variations occur in the atmosphere particularly in temperate regions on a global basis.

Pages: 215

4. How is sulfur dioxide produced and why is it important?

Answer: Sulfur dioxide is produced by burning coal and other fuels. Once in the atmosphere it dissolves in rain and causes acid rain. In addition, it combines with nitrous oxide from auto exhausts and forms additional acids. Acid deposition results in damaged vegetation, acidified lakes, and corrosion of metal and stone structures.

Pages: 212

5. How does the phosphorus cycle operate and why is it important?

Answer: The phosphorus cycle is the only cycle without a gaseous phase. Phosphorus is released from rocks and the soil, is incorporated into organic materials, and released into the sea as waste. Phospherus enrichment of lakes and ponds is called eutrophication. It is important as a fertilizer and if added to streams, ponds, and lakes in excess it causes rampant algal growth and reduces water quality.

CHAPTER 13 Grassland to Tundra

Multiple-Choice

Choose the one alternative that best completes the statement or answers the question.

Pages: 226
1. An important determinant of the grasslands is
   A) 50 inches of rainfall per year.
   B) fire.
   C) low rate of water loss.
   D) trees.
   Answer: B

Pages: 226
2. Which of the following are NOT true.
   A) Grasslands have a thick layer of mulch.
   B) The animals of grasslands are grazers or burrowing animals.
   C) Grasslands are evergreen and grow in the winter.
   D) Grassland vegetation is well adapted to grazing.
   Answer: C

Pages: 233
3. In grasslands, grazing
   A) increases the photosynthetic rate.
   B) relocates nutrients.
   C) removes photosynthetically inefficient tissue.
   D) all of the above
   Answer: D

Pages: 231
4. The strata of grasslands is composed of
   A) roots and herbaceous layers.
   B) a herbaceous layer only.
   C) a ground layer.
   D) roots, herbaceous, and ground layers.
   Answer: D

Pages: 231
5. Which of the following about grasslands in NOT true.
   A) Vegetation is seasonal, herbaceous parts die back.
   B) Roots are deep and extensive even more than in other communities.
   C) Requires a heavy rainfall (up to 40 inches per year).
   D) all of the above
   Answer: C

Pages: 227
6. The shortgrass prairie
   A) extends from desert grasslands east to the mixed grasslands of central U.S.
   B) has largely buffalo grass and blue grama grass.
   C) is a low moisture low humidity region.
   D) all of the above
   Answer: D

Pages: 235
7. Energy flow in a grassland results in
    A) about 50% of the herbage is consumed by small mammals.
    B) greater amounts of the primary production consumed below ground.
    C) invertebrate consumers are highly efficient.
    D) all of the above
    Answer: B

Pages: 228
8. Tallgrass prairies
    A) are composed of bunch grass like buffalo grass.
    B) are composed grasses and trees.
    C) are found in a narrow elongated region running north to south in middle America.
    D) A and B are correct
    Answer: C

Pages: 237
9. Savannas
    A) occur in mountainous areas on the dry side.
    B) occur in cool climates.
    C) have a low nutrient pool because of dryness.
    D) are composed of plants easily eliminated by fire.
    Answer: C

Pages: 237
10. Savannas have
    A) large numbers of herbivores.
    B) no invertebrates.
    C) no trees or shrubs.
    D) high rainfall (40 inches).
    Answer: A

Pages: 237
11. The animals commonly found in savannas
    A) include invertebrates.
    B) include animals that run fast.
    C) include ungulates.
    D) all of the above
    Answer: D

Pages: 240
12. Shrublands are composed of
    A) plants with woody, persistent short stems, and no central trunk.
    B) communities of grass and small trees.
    C) desert and grass.
    D) desert and small trees.
    Answer: A

Pages: 241
13. In the U.S. the mediterranean-type shrubland community is found in
    A) Oregon.
    B) California.
    C) New York.
    D) Alabama.
    Answer: B

Pages: 241
14. In North America
    A) the shrubland community is called Chaparral.
    B) many shrubs require fire for the seeds to germinate.
    C) shrublands are of the Mediterranean type.
    D) all of the above
    Answer: D

Pages: 246
15. Land where evaporation exceeds rainfall is called
    A) shrubland.
    B) grassland.
    C) desert.
    D) tundra.
    Answer: C

Pages: 247
16. Desert plants are
    A) grasses.
    B) small trees.
    C) woody-stemmed and soft brittle-stemmed shrubs.
    D) soil algae.
    Answer: C

Pages: 249
17. Nutrient cycles in deserts
    A) do not cycle well.
    B) have excess nitrogen.
    C) have excess potassium.
    D) have the nutrient supply in the soil.
    Answer: A

Pages: 250
18. Tundra has
    A) low temperatures.
    B) heavy rainfall.
    C) high species diversity.
    D) all of the above
    Answer: A

Pages: 251
19. The permafrost region of the soil in the arctic tundra
    A) does not melt in the summer.
    B) because of the permafrost, water won't penetrate and the tundra is very wet.
    C) because of the permafrost, plants can live in a very low rainfall area.
    D) all of the above
    Answer: D

Pages: 256
20. Because of its coldness and high moisture, the tundra has
    A) abundant amphibians.
    B) abundant reptiles.
    C) abundant insect but few species.
    D) no mammals.
    Answer: C

True-False

Write T if the statement is true and F if the statement is false.

Pages: 226
1. Grasses are either sod formers or bunch grass.
   Answer: True

Pages: 226
2. Grasslands account for 50% of the earth's surface.
   Answer: False

Pages: 228
3. Shortgrass prairies have been damaged by the plow.
   Answer: False

Pages: 234
4. Grazing stimulates primary growth in the prairie regions.
   Answer: True

Pages: 235
5. Most of the primary production in grassland is above ground.
   Answer: False

Pages: 236
6. The accumulation of mulch in the prairie can have a detrimental effect on nitrogen cycling.
   Answer: True

Pages: 228
7. For the most part, the tallgrass prairies are no more having been converted to corn and other cropland.
   Answer: True

Pages: 237
8. Unlike prairies, fires are not a determinant for savannas.
   Answer: False

Pages: 231

9. Grasses with extensive underground roots are well adapted to fire, drought, and grazing.
   Answer: True

Pages: 228

10. North American grasslands consist of six vegetation types: tallgrass, shortgrass, mixed prairies, palouse prairies, annual grassland, and desert grassland.
    Answer: True

Pages: 232

11. Prairie have three strata: ground layer, herbaceous layer, and woody layer.
    Answer: False

Pages: 232

12. One of the most common mammals of the prairies is the lemming.
    Answer: False

Pages: 231

13. The ground layer of the prairies consists of shortgrass.
    Answer: False

Pages: 251

14. Tundra has a permanantly frozen soil called the permafrost.
    Answer: True

Pages: 241

15. The chaparral of California is a special grassland with small shrubs.
    Answer: False

Pages: 247

16. Dry areas due to lack of rainfall commonly occur on the lee side of mountains. The community here is a desert community.
    Answer: True

Pages: 247

17. In deserts, the root mass exceeds the shoot mass.
    Answer: False

Pages: 250

18. Tundra is a community shaped by low temperatures, short growing season, and low rainfall.
    Answer: True

Pages: 255

19. Alpine tundra has a thicker permafrost than arctic permafrost.
    Answer: False

Pages: 252

20. Tundra vegetation is perennial and reproduces vegetatively.
    Answer: True

Pages: 255

21. In tundras, the major carnivore is the wolf which preys on lemmings, musk ox, and caribou.
    Answer: True

Short Answer

Write the word or phrase that best completes each statement or answers the question.

Pages: 226
1. _____ occur in regions of 25 - 75 cm/yr, of rainfall too light to support forest growth and too heavy to be a desert.
Answer: Grasslands

Pages: 226
2. Wheat, oats, barley, and corn are commonly cultivated in former _____.
Answer: grasslands

Pages: 226
3. The two basic types of grasses are _____ and _____.
Answer: sod grass, bunch grass   (any order)

Pages: 227
4. _____ grasslands are found in forested regions.
Answer: Successional

Pages: 230
5. The grasslands of South America are the _____.
Answer: pampas

Pages: 227
6. The five major types of grasslands (prairies) are: _____ (name all five).
Answer: tallgrass, mixed, shortgrass, palouse, annual
          (any order)

Pages: 232
7. Two dominant animal groups of the prairie are _____ and _____ animals.
Answer: ungulates, burrowing   (any order)

Pages: 232
8. Grassland animals share two outstanding traits: The ability to _____ and to _____ fast.
Answer: leap, run (in order)

Pages: 233
9. In the prairies, the _____ layer is more pronounced than in any other major ecosystem.
Answer: root

Pages: 234
10. Numerous studies have shown the moderate grazing _____ primary production.
Answer: increases

Pages: 236
11. A prairie in tropical areas with woody plants, found in Africa, is the _____.
Answer: savanna

Pages: 237
12. Savannas support a large and varied group of _____, both _____ and _____ because of the grass and woody species.
Answer: herbivores, grazing, browsing   (last two any order)

Pages: 239
13. Covering large portions of the arid and semi-arid world is a climax woody community called a/an

_____.
Answer: shrubland

Pages: 244
14. Shrub ecosystems, successional, or mature, are characterized by small _____ plants and increased stratification over grasslands.
Answer: woody

Pages: 246
15. A climax community where evaporation exceeds rainfall is call a/an _____.
Answer: desert

Pages: 247
16. _____ and _____ shrubs, yucca, cacti, and small trees are characteristics of the desert vegetation.
Answer: woody stemmed, soft brittle stemmed   (any order)

Pages: 249
17. Small animals that feed on seeds are called _____.
Answer: granivores

Pages: 250
18. The community found in cold regions and named the "treeless plain" by the finnish people is the

_____.
Answer: tundra

Pages: 252
19. The two types of tundra _____ and _____ both occur in cold regions and have comparable but different plants and animals.
Answer: arctic, alpine   (any order)

Pages: 251
20. Arctic tundras have a/an _____ under the soil.
Answer: permafrost

Pages: 251
21. Daily primary production rates of the tundra of 0.9 - 1.9 gm/square meter are _____ to prairies.
Answer: comparable

Matching

Choose the item from Column 2 that best matches each item in Column 1.

Pages: 246
1. Column 1: High daily temperatures and cold nights with very low precipitation
   Column 2: desert

Pages: 250
2. Column 1: Cold temperature most of the year with sparse precipitation
   Column 2: tundra

Pages: 237
3. Column 1: High temperature, precipitation exhibits extreme seasonal fluctuations about a moderate amount.
   Column 2: savanna

Pages: 239
4. Column 1: Warm daily temperatures to hot with minimal precipitation.
   Column 2: shrubland

Pages: 226
5. Column 1: Warm to hot and too dry to support support forest growth
   Column 2: prairie

Essay

Write your answer in the space provided or on a separate sheet of paper.

Pages: 226
1. What conditions lead to a grassland community such as a tallgrass prairie?
   Answer: Prairies are formed in climatic areas of moderate rainfall where it is too dry for tree growth. Such prairies can form with bunch grass and sod forming species as well as some forbes. Fire is a major determinant as it kills trees but not the deep roots of prairie plants.

Pages: 233
2. What are some characteristics of grassland animals and why are they important?
   Answer: Grassland animals are either grazing or burrowing animals. Burrowing provides shelter and protection from predators. Grazing animals lack shelter and are capable of running fast. The invertebrates are also important as they are responsible for some breakdown of dead grass.

Pages: 240
3. How do the abiotic components affect the plants of the shrublands?
   Answer: Shrublands are distributed in arid to semi-arid and in some temperate regions, moisture plays a major role in the types and amounts of vegetation. The shrublands are highy variable including heathlands and chaparral. Most shrubs are xeric, evergreen, with woody persistent stem and no central trunk.

Pages: 228

4. What impact have humans had on the shortgrass and tallgrass prairies?

    Answer: Humans have had a devastating effect on the prairies. First, the dense sod was plowed and broken up and then cattle overgrazed the land. The result was the dustbowl of the 1930's and the destruction of major areas for years to come.

Pages: 236

5. Describe the nitrogen cycle in grasslands.

    Answer: About 90% of the nitrogen in grasslands is tied up in soil organic matter, about 2% in litter, and about 5% in live and dead plant cover. Turnover of nitrogen is fairly rapid and most of the nitrogen that enters a green plant one growing season will reenter another the following growing season. Mulch can inhibit nitrogen cycling by intercepting nitrogen in rainfall and inhibiting nitrogen fixation by nitrogen-fixing microbes. Periodic fires clear away accumulated mulch and stimulated the growth of nitrogen-fixing legumes.

Pages: 236

6. How do savannas differ from grasslands?

    Answer: Savannas develop under conditions of high but brief rainfall each year. The rainfall is seasonal but temperatures are warm. As a result of this, growth of trees is supported as well as grassland. In the grasslands rainfall occurs throughout the year with dry spells and cold winters. Savannas do not have a cold winter. Fire plays an important role in each.

CHAPTER 14 Forests

Multiple-Choice

Choose the one alternative that best completes the statement or answers the question.

Pages: 283
1. Which of the following is NOT a temperate forest?
   A) deciduous forest
   B) montane forest
   C) tropical rain forest
   D) coniferous forest
   Answer: C

Pages: 265
2. Along the Northwest coast of the U.S. is an area of dense summer fog, high rainfall, and moist air. The forest in this area is
   A) temperate rain forest.
   B) montane forest.
   C) taiga.
   D) deciduous forest.
   Answer: A

Pages: 264
3. The main boreal forest has
   A) continuous stands of coniferous trees, moss, and low shrubs.
   B) oaks and maple in disturbed areas.
   C) has permafrost in southern areas with mixed hardwoods.
   D) occurs in the southern U.S.
   Answer: A

Pages: 268
4. Which of the following is NOT found in coniferous forests?
   A) pines with straight cylindrical trunks
   B) dense to thin crowns depending on species
   C) deciduous conifers
   D) deciduous broad leaved trees
   Answer: D

Pages: 268
5. Vertical stratification in a coniferous forest
   A) is well developed.
   B) is not well developed.
   C) has a ground layer where nothing grows.
   D) has an immense array of small plants in the understory.
   Answer: B

Pages: 268

6. In coniferous forests,
    A) insect populations are large and frequently destructive to trees.
    B) have an extensive mammal species unique to the coniferous forest.
    C) earthworm species are numerous.
    D) very little nitrogen accumulates in the pine trees.
    Answer: A

Pages: 269

7. In coniferous forests, nitrogen and minerals
    A) are cycled rapidly as soils are thick.
    B) are internally cycled.
    C) are not cycled at all.
    D) do not accumulate in litter.
    Answer: B

Pages: 273

8. Deciduous forests covered much of eastern U.S. but they are highly fragmented because of:
    A) logging.
    B) farming and clearing of the land.
    C) foreign diseases and insects.
    D) all of the above
    Answer: D

Pages: 271

9. Old growth Douglas fir supports the growth of cyanophycophyllous lichens which
    A) are consumers.
    B) are decomposers.
    C) fix nitrogen.
    D) accumulate calcium.
    Answer: C

Pages: 273

10. Which of the following is not strictly a broadleafed forest but contains conifers also?
    A) oak-hickory
    B) northern hardwoods
    C) mixed mesophytic
    D) temperate evergreen forests.
    Answer: D

Pages: 276

11. In the highly developed uneven aged deciduous forests, there are
    A) two strata: canopy and soil.
    B) three strata: canopy, litter, and soil.
    C) there is basically one stratum because the soil is thin.
    D) four strata: canopy, understory, shrubs, and ground cover.
    Answer: D

Pages: 279

12. Deciduous forests
    A) cycle large amounts of nitrogen.
    B) cycle less minerals than coniferous forests.
    C) litter layer is not an important nutrient pool.
    D) do not cycle minerals.
    Answer: A

Pages: 280

13. Tropical forests are the most diverse of the forest types and include
    A) rain and cloud forests.
    B) seasonal forests.
    C) dry forests.
    D) all of the above
    Answer: D

Pages: 283

14. In the tropical rain forest, highest strata are
    A) trees which are very tall, 60-80 meters.
    B) shrubs and small trees.
    C) climbers which are vinelike.
    D) low stunted trees.
    Answer: A

Pages: 284

15. The most numerous animals in a tropical rain forest are
    A) mammals.
    B) amphibians.
    C) insects.
    D) birds.
    Answer: C

True-False

Write T if the statement is true and F if the statement is false.

Pages: 264
1. The taiga is a conifer forest.
   Answer: True

Pages: 264
2. The boreal forest is composed of redwoods, hemlock, and lodgepole pine.
   Answer: False

Pages: 264
3. In a boreal forest, Picea, Abies, Pinus, Larix are the principal conifer species.
   Answer: True

Pages: 268
4. In coniferous forests because of high crown density and deep shade, the lower strata are well developed.
   Answer: False

5. Very little above ground biomass is found in coniferous forests.
    Answer: False

6. In coniferous forests, decomposition is slow and litter accumulates.
    Answer: True

7. Seasonal fluctuation in light is a chief characteristic of the coniferous forest.
    Answer: False

8. Insects are very important in the coniferous forest and many attack the trees but earthworms are few.
    Answer: True

9. Coniferous forests cycle huge quantities of nutrients but do not accumulate any.
    Answer: False

10. In deciduous forests, the most rapidly available nutrients are in the litter.
    Answer: True

11. In the deciduous forest, the uppermost canopy is the brightest with light diminishing as it goes downward.
    Answer: True

12. The species composition of the broad leaved forests has been significantly altered by human activities.
    Answer: True

13. A large standing biomass is characteristic of the tropical rain forest.
    Answer: True

14. Trees in the tropical rain forest are unique because they do not support the growth of mycorrhizae.
    Answer: False

Short Answer

Write the word or phrase that best completes each statement or answers the question.

1. Temperate forests are of two types: the _____ forest and the _____ forest.
    Answer: deciduous, coniferous   (any order)

Pages: 264

2. All forests possess large amounts of _____.
   Answer: aboveground biomass

Pages: 264

3. Old forests, products of evolution with complex species interactions have been largely replaced by less complex _____.
   Answer: secondary forests

Pages: 264

4. The most northern of the coniferous forests is the _____.
   Answer: taiga or boreal forest

Pages: 266

5. The four coniferous evergreen forests in the U.S. are _____ (name all four).
   Answer: taiga, temperate rain forest, montane coniferous forest, southern pine forest (any order)

Pages: 273

6. Temperate broadleaved forests have trees that _____ their leaves in the fall and are dormant in the winter.
   Answer: lose

Pages: 270

7. Nutrient cycling in coniferous forests typically is characterized by a relatively high proportion of nutrient sequestered in foliage and woody biomass withdrawn from short-term cycling. For this reason conifers are often called _____.
   Answer: accumulators

Pages: 265

8. _____ of Engelmann spruce, fir, and Ponderosa pine are found in the Sierra Nevada, Rocky Mountains, and Cascades.
   Answer: Montane forests

Pages: 270

9. In a _____ forest, the ground layer plants consist mostly of lichens and moss.
   Answer: spruce

Pages: 264

10. The main boreal forest has two broadleaved genera _____ and _____.
    Answer: Populus, Betula   (any order)

Pages: 270

11. Uneven-aged broadleaved forests usually have the following _____ strata: _____ (name all).
    Answer: four, upper canopy, lower canopy, shrub layer, ground layer
    (last four any order)

Pages: 280

12. Nutrient studies of coniferous forests and deciduous forests have shown that _____ nutrients are cycled through deciduous forests than through coniferous forests.
    Answer: more

Pages: 283
13. There are _____ general strata in a tropical rain forest but _____ is poorly defined.
    Answer: six, stratification   (in order)

Pages: 283
14. A significant component of the tropical rain forest are _____, plants that depend on trees for support.
    Answer: epiphytes

Pages: 287
15. Recent destruction of the tropical rain forest has eliminated the high _____ of _____ and _____ life.
    Answer: diversity, plant, animal   (last two any order)

Matching

Choose the item from Column 2 that best matches each item in Column 1.

Pages: 277
1. Column 1: Trees are bare during the winter reducing water loss.
   Column 2: temperate deciduous forest

Pages: 265
2. Column 1: In addition to trees, the glaciated region is an area of lakes and bogs.
   Column 2: boreal forest

Pages: 273
3. Column 1: Largely replace by agriculture in Asia and Europe.
   Column 2: temperate deciduous forest

Pages: 283
4. Column 1: Has five general strata.
   Column 2: tropical rain forest

Pages: 283
5. Column 1: Most productive of the natural communities.
   Column 2: tropical rain forest

Essay

Write your answer in the space provided or on a separate sheet of paper.

Pages: 285
1. Tropical rain forests are considered to be the most productive communities on earth. Why?
   Answer: Tropical rain forests are highly productive because of concentration and rapid decomposition of organic matter on forest floor, rapid recycling of nutrients, highly developed mycorrhizae, warm temperature, heavy amounts of precipitation, and long growing season.

Pages: 278
2. What are the differences in nutrient cycling between a boreal forest and a temperate broad leaf forest?

   Answer: The boreal forest processes relatively little nutrients in the biogeochemical cycle. Nitrogen and other nutrients are conserved in the leaves and trunk and there is very little in the soil. Temperate broadleaf forests lose leaves yearly. They break down quickly, so larger quantities of nitrogen and other elements are cycled.

Pages: 268,283
3. What is the role of stratification in the coniferous forest and the tropical rain forest?

   Answer: Stratification indicates diversification as more habitats are formed. The coniferous forest has at most poorly developed stratification whereas the tropical forest has five general layers. As a consequence, there is more diversity of life in the tropical rain forest.

Pages: 286
4. What is the function of mycorrhizae?

   Answer: Mycorrhizae are fungi growing in or around the roots of plants. They increase mineral uptake of host plant and because they attach to dead material directly, transfer nutrients to roots of plants.

Pages: 284
5. How does stratification in the tropical rain forest influence animal life?

   Answer: The five strata in the tropical rainforest provide many animals with specific habitats. Each stratum supports its own array of animal life. The forest floor is dominated by termites and ants, along with such herbivores as deer. The upper canopy is home to primates and bats.

Multiple-Choice

Choose the one alternative that best completes the statement or answers the question.

Pages: 296
1. During the spring and fall overturn
   A) plankton are carried to lower zones.
   B) nekton are unaffected.
   C) no material can accumulate in the benthos due to the turbulence.
   D) all of the above
   Answer: A

Pages: 296
2. The most abundant organisms in the benthos region are
   A) periphyton.
   B) nekton.
   C) anaerobic bacteria.
   D) rooted plants.
   Answer: C

Pages: 297
3. Primary production in lentic systems
   A) are carried out by nekton.
   B) are carried out in the limentic zone by phytoplankton.
   C) occur in the benthic region.
   D) are not dependent on nutrients.
   Answer: B

Pages: 295
4. Significant photosynthesis by macrophytes occurs in
   A) the littoral zone.
   B) the hypolimnion.
   C) the benthos.
   D) the tropholytic zone.
   Answer: A

Pages: 294
5. Fish and other free swimming organisms are
   A) macrophytes.
   B) plankton.
   C) zooplankton.
   D) nekton.
   Answer: D

Pages: 290
6. Lentic ecosystems
   A) are waterfalls.
   B) are streams.
   C) are lakes and ponds.
   D) are dry areas in the ocean.
   Answer: C

Pages: 294
7. Phytoplankton are
   A) eaten by zooplankton.
   B) found in the benthos.
   C) not primary producers.
   D) are emergents.
   Answer: A

Pages: 301
8. Excessive nutrients from raw sewage or agricultural lands causes lakes to become
   A) oligotrophic.
   B) dystrophic.
   C) eutrophic.
   D) a marl lake.
   Answer: C

Pages: 302
9. The _____ of the current molds the stream.
   A) resistance
   B) velocity
   C) temperature
   D) all of the above
   Answer: B

Pages: 303
10. The aufwuchs consist of
    A) macrophytes.
    B) fish.
    C) diatoms, algae, etc.
    D) aquatic worms.
    Answer: C

Pages: 298
11. A eutrophic lake
    A) is high in nitrogen and phosphorus.
    B) is low in nutrients.
    C) has no phytoplankton.
    D) all of the above
    Answer: A

Pages: 305
12. A major source of energy and nutrients is
    A) detrital material from outside the stream.
    B) CPOM.
    C) FPOM.
    D) all of the above
    Answer: D

Pages: 308
13. The first organisms to breakdown CPOM are
    A) bacteria.
    B) shredders.
    C) scrapers.
    D) all of the above
    Answer: A

Pages: 309
14. Nutrient cycling in lotic systems
    A) spirals downward by cycling and transport.
    B) stay in one place.
    C) move slowly.
    D) does not take place, the currents are too fast.
    Answer: A

Pages: 310
15. In a river continuum
    A) there is an increase in heterotrophy as it goes downstream.
    B) the organisms stay the same.
    C) shredders become increasingly important.
    D) streams go from heterotrophy to autotrophy as they proceed downstream.
    Answer: D

Pages: 311
16. The source of nutrients for bottom dwelling organisms at the mouth of a river is
    A) dissolved organic matter.
    B) drift and small particles.
    C) phytoplankton.
    D) nekton.
    Answer: B

Pages: 313
17. Dams with a large pool of water become
    A) stratified.
    B) more like the lotic environment producing them.
    C) stagnant.
    D) none of the above
    Answer: A

Pages: 313
18. Lands halfway between aquatic and terrestrial systems are
    A) drained lakes.
    B) filled in rivers.
    C) wetlands.
    D) ridge tops.
    Answer: C

Pages: 322
19. The animal life in a wetland is predominantly
    A) herbivores.
    B) carnivores.
    C) predators.
    D) decomposers.
    Answer: A

Pages: 322
20. The major nitrogen source in a peatland is
    A) water.
    B) bacteria.
    C) carnivorous plants like sundew.
    D) the atmosphere.
    Answer: D

True-False

Write T if the statement is true and F if the statement is false.

Pages: 290
1. The temperature gradient in a pond is the thermocline.
   Answer: True

Pages: 290
2. The metalimnion is in the benthos region.
   Answer: False

Pages: 292
3. In aquatic habitats, oxygen is an important limiting factor.
   Answer: True

Pages: 293
4. Carbon dioxide in the lentic community is in the form of carbonic acid.
   Answer: True

Pages: 293
5. The trophogenic zone refers to the hypolimnion.
   Answer: False

6. Lentic ecosystems have a littoral zone in which oxygen is absent and anaerobic bacteria predominate.
   Answer: True

Pages: 294
7. The limnetic zone is inhabited by rooted plants such as water lilies.
   Answer: False

Pages: 294
8. Free-swimming organisms such as fish are called the nekton.
   Answer: True

Pages: 294
9. The benthos refers to the bottom of a lake.
Answer: True

Pages: 297
10. Primary production is the accumulation of detritus from the surrounding area.
Answer: False

Pages: 298
11. Oligotrophic refers to streams with excess nutrients.
Answer: False

Pages: 301
12. The amount of oxygen required to oxidize organic matter in a stream is the BOD.
Answer: True

Pages: 302
13. Velocity of the current molds the stream.
Answer: True

Pages: 304
14. Primary production from algae is the chief energy source in lotic ecosystems.
Answer: False

Pages: 306
15. In lotic environments, bacteria are associated with FPOM.
Answer: True

Pages: 308
16. Shredders such as caddisfly larvae feed on the DOM.
Answer: False

Pages: 312
17. Lotic communities change along a continuum from small streams which are primarily heterotrophic to large streams which are more autotrophic.
Answer: True

Pages: 313
18. Waterlogged soil, deep water, and trees define a wetland.
Answer: False

Pages: 315
19. Marshes, fens, sounds, and bays are some typical wetlands.
Answer: True

Pages: 318
20. The most important characteristic of wetland is its hydroperiod.
Answer: True

Short Answer

Write the word or phrase that best completes each statement or answers the question.

Pages: 289
1. Freshwater ecosystems fall into two categories, _____ ecosystems contained in a basin or lake and _____ systems consist of flowing water.
   Answer: lentic, lotic   (in order)

Pages: 290
2. The study of freshwater systems is known as _____.
   Answer: limnology

Pages: 290
3. The upper warmer layer in a lake is the _____.
   Answer: epilimnion

Pages: 290
4. The area of rapid change from the warm surface layer to a cool bottom layer in a lake is the

   _____.
   Answer: thermocline

Pages: 290
5. The cool bottom layer of a lake is the _____.
   Answer: hypolimnion

Pages: 294
6. The _____ zone is comprised of rooted plants along the edge of the lake where light penetrates to the bottom.
   Answer: littoral

Pages: 293
7. Under ice _____ may become limiting and kill fish.
   Answer: oxygen

Pages: 284
8. The _____ zone or open water is inhabited by _____, free swimming organisms and _____, suspended organisms.
   Answer: limnetic, nekton, plankton   (in order)

Pages: 296
9. The dominant organisms of the _____ zone are anaerobic bacteria.
   Answer: benthic

Pages: 298
10. Nutrient enriched lakes are _____ lakes and nutrient poor lakes are _____.
    Answer: eutrophic, oligotrophic   (in order)

Pages: 301
11. When the oxygen available is not sufficient to oxidize the organic matter in a body of water, the amount needed is called the _____.
    Answer: biochemical oxygen demand (BOD)

Pages: 303
12. The nutrients for a flowing water community come from the _____ that supplies it with water.
    Answer: watershed

Pages: 309
13. Streams carry _____ and _____ as very little primary production takes place.
    Answer: dissolved organic matter (DOM), particulate organic matter (POM)   (any order)

Pages: 308
14. _____ feed on leaves; they are invertebrates such as craneflies, caddisflies, and stoneflies.
    Answer: Shredders

Pages: 308
15. _____ is the chief energy source in a lotic environment.
    Answer: Detritus

Pages: 302
16. _____ of current determines the nature of lotic communities.
    Answer: Velocity

Pages: 308
17. Most fish are _____.
    Answer: predators

Pages: 314
18. In defining a wetland, _____, _____, and _____ must be considered.
    Answer: soil, vegetation, water conditions (any order)

Pages: 315
19. Some wetland types are _____ (five answers).
    Answer: marshes, swamps, bogs, sounds, bays (any order)

Pages: 315
20. An important feature of wetlands is the _____, which refers to the duration, frequency, depth, and season of flooding.
    Answer: hydroperiod

Matching

Choose the item from Column 2 that best matches each item in Column 1.

Pages: 290
1. Column 1: Nekton can be found in the epilimnion.
   Column 2: lake

Pages: 315
2. Column 1: Sphagnum and sedges are important primary producers.
   Column 2: bog

Pages: 303
3. Column 1: Riffles are the sites of primary production.
   Column 2: river

Pages: 316
4. Column 1: Forested wetlands.
   Column 2: swamp

Pages: 294
5. Column 1: Contains a littoral area of rooted plants.
   Column 2: lake

Essay

Write your answer in the space provided or on a separate sheet of paper.

Pages: 297,310
1. Compare nutrient cycling in a lentic ecosystem to nutrient cycling in a lotic ecosystem.
   Answer: <u>Lentic:</u> Nutrient cycling is somewhat self-contained; cycling occurs between water column and sediments with uptake by phytoplankton, zooplankton, bacteria and other consumers, and rooted aquatics. Rooted aquatic vegetation returns organic matter to sediments where decomposition takes place. Other inputs of nutrients come from precipitation and terrestrial detrital material blown or carried into lake. <u>Lotic or running water systems:</u> Nutrient cycling depends largely upon inputs of detrital material from terrestrial sources. Currents carry nutrients rapidly downstream. (See text pages.)

Pages: 290
2. What is the importance of the spring and fall overturn to the lentic community?
   Answer: The spring and fall overturns mix the nutrients, oxygen, and carbon dioxide and create a uniform distribution of water temperature.

Pages: 321
3. Although wetlands are named in relation to the plant life, the animal life is also important. What animals are present and what is their role?
   Answer: Herbivores such as microcrustaceans abound and consume the algae. Snails are present on the litter and geese, coots, mallards, and other waterfowl are typical inhabitants. Muskrats and rabbits are major herbivores. Racoon, fox, weasel, skunk, are typical predators.

Pages: 218
4. What is the importance of the hydroperiod to a wetland community?
   Answer: Hydroperiod refers to the duration, frequency, depth, and season of flooding. These factors determine the three broad classes and the characteristics of wetland communities. Based on hydroperiod there are basin wetlands, riverine wetlands, and fringe wetlands (near coast). Periods of drought can induce vegetational changes as can depth of water.

Pages: 322,310
5. Compare energy flow in a wetland with energy flow in a river.

    Answer: In a river the major source of energy comes from detrital input and downstream section depend upon energy input from upstream sources. Large rivers gain energy from some autotrophic production by rooted aquatics and phytoplankton. In wetlands the energy source is autotrophic. Energy flow, however, varies between types of wetland and their hydroperiod. In peat bogs decomposition is impaired because of poor decomposition and much of the energy accumulates in undecomposed organic matter, unavailable for further use in the system.

CHAPTER 16 Saltwater Ecosystems

Multiple-Choice

Choose the one alternative that best completes the statement or answers the question.

Pages: 326
1. The seas exhibit vertical stratafication as follows:
    A) pelagic and benthic regions
    B) photic and nearphotic regions
    C) trench and littoral regions
    D) estuaries and trenches
    Answer: A

Pages: 326
2. Vertically, oceans are sharp zones of temperature, light, and salinity called
    A) the photic zone.
    B) the mesopelagic zone.
    C) the bathypelagic zone.
    D) all of the above
    Answer: D

Pages: 326
3. Water in the oceanic trenches is called
    A) abyssal pelagic.
    B) neritic.
    C) hadalpelagic.
    D) littoral.
    Answer: C

Pages: 326
4. Lying over the major plains of the ocean and down to about 6000m is the
    A) neritic province.
    B) hadalpelagic.
    C) abyssal pelagic.
    D) none of the above
    Answer: C

Pages: 327
5. Oceans have a strong permanent thermocline in
    A) arctic regions.
    B) temperate regions.
    C) tropical regions.
    D) antarctic regions.
    Answer: C

Pages: 327

6. _____ make up 86% of sea salts.
    A) Sodium
    B) Chloride
    C) Calcium
    D) A and B
    E) A,B, and C
    Answer: D

Pages: 327

7. Sea water is a weakly buffered solution at
    A) pH 6.0-6.5.
    B) pH 3.0-6.0.
    C) pH 9.0-9.5.
    D) pH 8.0-8.3.
    Answer: D

Pages: 328

8. Salinity like temperature exhibits a vertical gradient in the ocean. It is called the
    A) pycnocline.
    B) halocline.
    C) osmocline.
    D) calcium gradient.
    Answer: B

Pages: 329

9. Langmuir cells
    A) are air currents over the ocean.
    B) are deep currents in the ocean abyssal region.
    C) are vertical currents from the surface to 200 meters caused by ocean currents.
    D) are circulating vertical cells spinning water just below the surface of the ocean.
    Answer: D

Pages: 329

10. Periodic and predictable rising and falling of the sea levels over an interval are
    A) Langmir cells.
    B) tides.
    C) Ekman spiraling.
    D) upwellings.
    Answer: B

Pages: 331

11. Primary producers in relatively unproductive seas is
    A) phytoplankton.
    B) nanoflagellates.
    C) cynobacteria.
    D) all of the above
    Answer: D

Pages: 331
12. The oceans are not as productive as the land because
    A) of not enough minerals.
    B) it is too cold.
    C) light does not penetrate to sufficient depths.
    D) all of the above
    Answer: C

Pages: 325
13. The most productive area of the ocean is/are
    A) the pelagic region.
    B) the abyssal region.
    C) the temperate oceans.
    D) are coastal waters and coral reefs.
    Answer: D

Pages: 331
14. Along rocky shores a pronounced variation occurs due to
    A) wave action and tides.
    B) seasonal temperature changes.
    C) low salt in water.
    D) all of the above
    Answer: A

Pages: 337
15. The black zone
    A) is an under water area.
    B) is paint and oil from ships washing up causing the black color.
    C) where water never reaches the granite rock.
    D) where Verrucaria type lichens and crustose algae grow.
    Answer: D

Pages: 337
16. In the littoral zone
    A) organisms are covered with water constantly.
    B) barnacles and mussels are common.
    C) ocean species are found.
    D) all of the above
    Answer: B

Pages: 340
17. The productivity of the kelp bed of the Pacific Northwest
    A) is less than a tropical rain forest.
    B) is greater than a tropical rain forest.
    C) is equal to a tropical rain forest.
    D) is heterotrophic only.
    Answer: B

Pages: 340
18. Tidal pools
    A) are distinct habitats.
    B) are subject to wide fluctuations in temperature.
    C) experience fluctuating salinity.
    D) all of the above
    Answer: D

Pages: 341
19. Life on sandy and muddy beaches has a characteristic infauna of
    A) nekton.
    B) barnacles.
    C) lugworms.
    D) fish.
    Answer: C

Pages: 343
20. The sandy beach
    A) has extensive producers.
    B) has energy supplied by detritus.
    C) does not cycle much energy.
    D) all of the above
    Answer: B

Pages: 344
21. Horseshoe-shaped coral reefs containing a lagoon formed from a volcano are
    A) atolls.
    B) barrier reefs.
    C) fringe reefs.
    D) none of the above
    Answer: A

Pages: 345
22. Coral reefs
    A) are not productive ecosystems.
    B) do not accumulate nor cycle many nutrients.
    C) are very productive with great diversity of species.
    D) are temporary cover.
    Answer: C

Pages: 346
23. The most important problems facing estuarian organisms are
    A) maintenance of position and adjusting to salinity.
    B) maintenance of oxygen level and salinity.
    C) maintenance of water coverage and carbon dioxide level.
    D) adjusting to salinity and carbon dioxide level.
    Answer: A

24. Tidal sands or mudflats developing into flat grasslands occur in temperate regions. These ecosystems are called
    A) salt marsh ecosystems.
    B) mangrove ecosystems.
    C) mudflat-sand ecosystems.
    D) all of the above
    Answer: A

25. Mangroves in mangrove wetlands
    A) have pneumatophores.
    B) have seeds germinating on the plant.
    C) grow in anoxic mud.
    D) all of the above
    Answer: D

True-False

Write T if the statement is true and F if the statement is false.

1. In the mesopelagic layer from 200 to 1000 meters deep in the ocean, large amounts of algae grow as light penetrates easily.
   Answer: False

2. The bottom of the sea is called the benthic region.
   Answer: True

3. Tropical oceans whose surface waters are constantly heated lack a thermocline.
   Answer: False

4. Sodium and chlorine make up less than 20% of the sea salts.
   Answer: False

5. Winds blowing across the surface of the ocean generate circulating vertical cells called Langmuir cells.
   Answer: True

6. Tides are random variations in level of the ocean influenced by the moon.
   Answer: False

7. Pelagic ecosystems have well defined communities due to light layers.
   Answer: False

Pages: 331
8. In regions of upwelling the dominant forms of phytoplankton are diatoms.
   Answer: True

Pages: 332
9. Nekton range in size from small fish to sharks and whales.
   Answer: True

Pages: 333
10. Benthic organisms are strictly autotrophic.
    Answer: False

Pages: 334
11. The oceans are more productive than terrestrial communities as evidence by nutrient depletion.
    Answer: False

Pages: 337
12. On rocky shores, the supralittoral fringe is permanently underwater and has ocean species.
    Answer: False

Pages: 337
13. Some animals found in the littoral zone of a rocky shore are barnacles, blue mussel, and limpets.
    Answer: True

Pages: 337
14. In the North Atlantic, giant kelp is found on rocky shores.
    Answer: False

Pages: 341
15. The sandy beaches and mud flats are largely autotrophic areas.
    Answer: False

Pages: 342
16. Burrowing animals would not be found on sandy beaches.
    Answer: False

Pages: 344
17. Fringing coral reefs grow from extinct volcanos.
    Answer: False

Pages: 347
18. Estuaries are autotrophic and heterotrophic, and organic matter must be washed in.
    Answer: True

Pages: 347
19. The organisms of an estuary are essentially marine and must be able to tolerate salt water.
    Answer: True

Pages: 352
20. Shallow depressions called salt flats, in the salt marsh community have edges which are colonized by salt tolerant plants such as <u>salicornia</u>.
    Answer: True

Short Answer

Write the word or phrase that best completes each statement or answers the question.

Pages: 326
1. The ocean is divided into two main layers, the _____: the whole body of water, and the _____ : the bottom.
   Answer: pelagic region, benthic region   (in order)

Pages: 326
2. The top layer of the pelagic region is the _____ zone which is about 200 meters deep and has sharp gradients in light, temperature, and salinity.
   Answer: photic

Pages: 326
3. The sea bottom is divided into three zones: _____(name all three).
   Answer: bathyl zone, abyssal zone, hadal zone   (any order)

Pages: 326
4. Within the pelagic region there are three layers: _____(name all three).
   Answer: photic zone, mesopelagic zone, bathylpelagic zone (any order)

Pages: 326
5. Horizontally the pelagic region is divided into three provinces: _____(name all three).
   Answer: neritic province, oceanic province, abyssal pelagic (any order)

Pages: 6
6. Elements commonly found in the sea include: _____.
   Answer: sodium, chlorine, sulfur, magnesium, potassium, calcium   (any order)

Pages: 329
7. Some characteristics of the ocean environment are: _____(name five).
   Answer: pressure, waves, Langmuir cells, Ekman spiralling, tides   (any order)

Pages: 331
8. The primary producers of the ocean are _____.
   Answer: phytoplankton

Pages: 332
9. _____ are free swimming organisms which feed on zooplankton and include small fish and sharks.
   Answer: Nekton

Pages: 333
10. _____ found on the bottom near volcanic areas spew forth hot water and minerals.
   Answer: Hydrothermal vents

Pages: 335
11. Zones of high production are _____(name three).
   Answer: coastal waters, estuaries, upwellings (any order)

12. The four major zones in the rocky shore ecosystem are _____ (name all four).
    Answer: supralittoral fringe, littoral zone, infralittoral, sublittoral   (any order)

13. Zonation on rocky shores or the North Atlantic results in an infralittoral fringe containing the brown algae known as _____.
    Answer: kelp (laminaria)

14. The two major animal groups on sandy shores and mudflats are _____ and _____.
    Answer: infauna, epifauna   (any order)

15. On sandy shores and mudflats low oxygen due to bacterial action, causes a layer of ferrous sulfides to form a _____.
    Answer: black layer

16. Coral reefs are complex associations of coral in the gastrodermal layer where the algae _____ live.
    Answer: zooxanthellae

17. Energy in salt marsh ecosystems is supplied by _____ which is formed from the salt grass _____.
    Answer: detritus, Spartinia (in order)

18. Where salt and freshwater meet a/an _____ is formed.
    Answer: estuary

19. Adults of the _____ bass live in the marine environment but the young fish grow up in the estuary.
    Answer: striped

20. In tropical regions the tidal flats, where wave action is absent and sediments accumulate are occupied by _____.
    Answer: mangrove swamps

Matching

Choose the item from Column 2 that best matches each item in Column 1.

The rocks, sandy shores, and mudflats are exposed to air at intervals by the changing tides resulting in three zones. Answer the questions regarding these three zones.

1. Column 1: Beach fleas and sand colored ghost crabs occupy this region on sandy beaches.
   Column 2: supralittoral zone

Pages: 337
2. Column 1: The region on rocky shores where Laminaria grows.
   Column 2: infralittoral zone

Pages: 339
3. Column 1: Mussels grow in this zone on a rocky shore.
   Column 2: littoral zone

Pages: 343
4. Column 1: The zone on a sandy shore where true marine life appears.
   Column 2: littoral zone

Pages: 343
5. Column 1: Starfish and sand dollars are found in this zone on a sandy beach.
   Column 2: littoral zone

Pages: 337
6. Column 1: Rocky shore with lichens and crustose algae.
   Column 2: supralittoral zone

Pages: 337
7. Column 1: Rocky shore where barnacles occur.
   Column 2: littoral zone

Essay

Write your answer in the space provided or on a separate sheet of paper.

Pages: 343
1. What is the importance of detritus to the food webs of estuaries, mud flats, and sandy beaches?
   Answer: In all cases the communities are primarily heterotrophic and detritus is the main
            source of organic material.

Pages: 351
2. Describe the zonation in a rocky shore, an estuary, a salt marsh, and a mangal. What causes
   zonation in each situation?
   Answer: The tide is the most important factor in causing zonation. There is a region which is
            under water for a short time only, a region under water more frequently and for a
            longer time, and a region under water most of the time. The organisms tending to be
            in one zone would be different from the organisms in other zones.

Pages: 350
3. In the salt marsh different species of the genus Spartinia are found in various habitats. What
   species are adapted to the different marsh regions?
   Answer: Spartinia alternifolia grows submerged in salt water at high tide whereas another
            species Spartinia patens is the salt meadow grass that grows in the high marsh
            flooded only at spring tides. These two species are adapted to two different
            environments of the salt marsh.

Pages: 331

4. Oceanic organisms of the pelagic region have some extreme conditions. How are plants and animals affected in the pelagic region?

   Answer: In the open sea there are no distinct communities because of the lack of supporting structures. In the photic zone, the primary producers are phytoplankton with zooplankton. The first herbivores and nekton feeding on both the phytoplankton and zooplankton.

Pages: 355

5. What conditions favor the development of mangals?

   Answer: In tropical regions, the mangrove wetlands replaces salt marshes on tidal flats. These occur where wave action is absent, sediments accumulate, and mud is anoxic. The mangrove roots have pneumatophores and the plants have prop roots.

CHAPTER 17 Properties of Populations

Multiple-Choice

Choose the one alternative that best completes the statement or answers the question.

Pages: 362
1. A population can be
   A) a group of demes.
   B) a genet.
   C) a ramet.
   D) an individual.
   Answer: A

Pages: 364
2. The density a population of plants can attain before crowding has some effect is the
   A) ecological effective density.
   B) density.
   C) crude density.
   D) ecological density.
   Answer: A

Pages: 364
3. In some areas, there are as many as 34 skunks per square mile. This number is the
   A) ecological effective density.
   B) ecological density.
   C) crude density.
   D) none of the above
   Answer: C

Pages: 366
4. The commonest distribution is
   A) random.
   B) clumped.
   C) even.
   D) in a line.
   Answer: B

Pages: 366
5. Uniform or regular distribution is usually obtained by
   A) scattering of seed.
   B) young staying near the mother until adulthood.
   C) territoriality.
   D) invertebrates of the forest floor.
   Answer: C

Pages: 367
6. Aggregations or clumping reflect
   A) a chance spreading.
   B) territoriality.
   C) some degree of interaction.
   D) all of the above
   Answer: C

Pages: 368
7. Regional distributions make up the _____ of a population.
    A) abundance
    B) grain
    C) ecological density
    D) range
    Answer: D

Pages: 368
8. Temporal dispersion
    A) can be circadian.
    B) can be lunar cycles.
    C) can be tidal cycles.
    D) all of the above
    Answer: D

Pages: 369
9. Which of the following is not a dispersal movement.
    A) circadian cycling of deer
    B) immigration
    C) emigration
    D) migration
    Answer: A

Pages: 372
10. The age structure of a population (age pyramids) is the
    A) abundance at some specific age.
    B) the frequency of a population in each age class vs. age class.
    C) relative density per year.
    D) all of the above
    Answer: B

Pages: 372
11. Theoretically all continuously breeding populations should tend toward a
    A) variable age distribution.
    B) higher frequency of individuals in older age classes.
    C) stable age distribution.
    D) decreasing population.
    Answer: C

Pages: 372
12. A ratio of 3:1 of young to adults is a/an
    A) normal distribution.
    B) aging population.
    C) stable population.
    D) growing population.
    Answer: D

Pages: 374
13. When studying age structures of forest trees _____ is used rather than age.
    A) height
    B) number of older trees
    C) basal area
    D) all of the above
    Answer: C

Pages: 374
14. Another characteristic of populations
    A) is the primary sex ratio.
    B) is the secondary sex ratio.
    C) is the older (adult) sex ratio.
    D) all of the above
    Answer: D

Pages: 376
15. Mortality is often expressed as
    A) crude death rate.
    B) one minus natality.
    C) probably of surviving.
    D) one minus the specific death rate.
    Answer: A

Pages: 376
16. A clear and systematic picture of mortality and survival in a population is best provided by
    A) a death rate table.
    B) a life table.
    C) specific birthrate.
    D) realized natality.
    Answer: B

Pages: 377
17. Life tables are calculated from
    A) life expectancy.
    B) survival rates.
    C) the number of individuals of a cohort counted each time period during the life span of the cohort.
    D) mortality rate.
    Answer: C

Pages: 379
18. Plant ecologists have adapted the life table to study population dynamics of
    A) shoot populations.
    B) modular growth.
    C) leaves.
    D) all of the above
    Answer: D

Pages: 380
19. A survivorship curve in which individuals tend to live out their physiological life span is
    A) typical of fish.
    B) typical of invertebrates.
    C) is typical of mammals.
    D) is not typical of any animals.
    Answer: C

Pages: 381
20. A survivorship curve which is linear is typical of
    A) fish.
    B) humans.
    C) clams.
    D) adult birds.
    Answer: D

Pages: 382
21. Most mortality curves
    A) are straight lines.
    B) are J shaped.
    C) curve upwards.
    D) are random.
    Answer: B

Pages: 384
22. Fecundity is
    A) the same as natality.
    B) the number of offspring produced per unit time by females in different age classes.
    C) the number of offspring produced per individual in an age class.
    D) the number of offspring in a population.
    Answer: C

True-False

Write T if the statement is true and F if the statement is false.

Pages: 362
1. Most populations know no boundaries other than those drawn by ecologists.
   Answer: True

Pages: 363
2. Plants are easy to study because it is necessary to sample only a few ramets.
   Answer: False

Pages: 364
3. The genes of a deme make up the gene pool of a population.
   Answer: False

Pages: 365
4. The number of mice per acre is a measure of the crude density.
   Answer: True

5. Clumped distributions are usually found in solitary animals.
   Answer: False

6. The distribution of humans is uniformally spaced because of housing.
   Answer: False

7. Regional distributions make up the range of the population.
   Answer: True

8. Migrations change dispersal movements as much as immigrations.
   Answer: False

9. Most natural populations are stable populations or reach this stage quickly.
   Answer: False

10. The age structure of a population refers to the ratio of individuals in a population of each age class for all age classes.
    Answer: True

11. The age structure diagram shows the frequency of prereproductive individuals on the bottom and the diagram is a pyramid in shape.
    Answer: True

12. The population of many third world countries, like India, has a very large proportion of individuals in the prereproductive class, therefore it is a declining population.
    Answer: False

13. When considering the age structure of trees, age is not a good indicator so ecologists use diameter at breast height or basal area.
    Answer: True

14. Life tables are used to interpret life expectancy and mortality.
    Answer: True

15. Survivorship curves are obtained by plotting the number of individuals of a particular age cohort vs. time.
    Answer: True

Short Answer

Write the word or phrase that best completes each statement or answers the question.

Pages: 362
1. A _____ is a group of interbreeding organisms of the same species which occupy a particular space at a particular time.
   Answer: population

Pages: 362
2. Populations consist of _____, a unit, or _____ a part of a larger unit.
   Answer: real individuals, modules (in order)

Pages: 363
3. Asexually reproducing individuals derived from a singe zygote are a/an _____, and several of these make up a _____. A single plant is a/an _____.
   Answer: ramet, clone, genet (in order)

Pages: 364
4. The sum of all genetic information carried by all members of a population is the _____.
   Answer: gene pool

Pages: 365
5. The number of individuals per unit area is the _____, but if calculated as the amount of area available as living space per individual it is the _____ density.
   Answer: crude density, ecological (in order)

Pages: 366
6. Patterns of dispersion are either _____ (name all).
   Answer: random, uniform, or clumped (any order)

Pages: 367
7. The moose is a very unsociable animal and shows a/an _____ distribution.
   Answer: random

Pages: 369
8. Dispersal may be one-way out of a habitat, _____
   Answer: emigration

Pages: 375
9. A graph of frequency vs age for age classes in a population is a/an _____ graph.
   Answer: age structure

Pages: 374
10. The _____ sex ratio occurs at conception and the _____ sex ratio is the sex ratio at birth.
    Answer: primary, secondary (in order)

Pages: 377
11. In a life table the _____ in years is usually calculated from cohort age and number alive at each year.
    Answer: mean expectation of further life or life expectancy

12. The survivorship curve is obtained by plotting the number of individuals in a/an _____ vs time.
   Answer: cohort

13. The survivorship curve has three periods: _____(name all three).
   Answer: dependency, reproductive, postreproductive (any order)

14. If high mortality rates occur in the dependency phase the graph curves _____ at the left and stays as it goes to the right.
   Answer: downward

15. Curves or tables of the appearence of new individuals are _____ or _____ curves.
   Answer: natality, fecundity (any order)

Matching

Choose the item from Column 2 that best matches each item in Column 1.

1. Column 1: Period of high mortality in type I survivorship curves.
   Column 2: post reproductive

2. Column 1: Period of high mortality in type III survivorship curves.
   Column 2: prereproductive (dependency)

3. Column 1: Most organisms survive this stage in a type I survivorship curve.
   Column 2: prereproductive (dependency)

4. Column 1: Rate of survivors is the same as other periods during this phase in a type II curve.
   Column 2: reproductive

5. Column 1: The age structure of an increasing population has the largest group in this class.
   Column 2: prereproductive (dependency)

6. Column 1: Large sized age groups in this age structure indicate a declining population.
   Column 2: post reproductive

Essay

Write your answer in the space provided or on a separate sheet of paper.

Pages: 366

1. What causes the three distributions: clumped, randomly, and uniformly?
   Answer: Behavior and habitat in animals are the primary causes. Clumping occurs when animals interact as in socially structured animals. Random distributions occur when there is no interaction. Even distributions occur with territorial animals.

Pages: 373

2. Plants require something other than age distributions to study populations. What is used and why?
   Answer: Module structures and asexual reproduction make an age distribution table impossible in plants. As a consequence diameter classes are used. Age and size studies have been done with leaves, stems, and in the case of cattail studies providing data on emergence, mortality, height, and growth of ramets.

Pages: 372, 376

3. Populations are described by many characteristics including density, age structure, and life tables. Define these terms.
   Answer: Density is usually referred to as crude density and is the number of individuals in a population at some time. Age distribution is the frequency of the population in each age group. It can indicate increasing, decreasing, or stationary size. The life tables use number of individuals of a cohort alive at each time interval to calculate the life expectancy of a cohort and survival of a cohort.

Pages: 368

4. Dispersal of populations is important to density considerations and population studies. What disperal types are there and how do they effect populations?
   Answer: Temporal dispersal is important to population density and age distributions. Loss of individuals by emigration lowers population size and age distribution could be changed if all were of one age class. Immigration increases density and sizes of classes. Migrations temporarily elimantes a population until the next year.

Pages: 384

5. Fecundity may be expressed as a net reproductive rate that combines survivorship with an age-specific schedule of births. What does this mean?
   Answer: Fecundity by definition is the number of offspring per individual per age class. In this sense fecundity is a net rate of reproduction and when plotted for various age classes suggests certain age classes contribute more to the population than others.

CHAPTER 18 Population Growth and Regulation

Multiple-Choice

Choose the one alternative that best completes the statement or answers the question.

Pages: 390
1. When growth of a population is calculated by using the age specific fecundity and the age specific survivorship a _____ is obtained.
   A) an exponential equation
   B) a linear equation
   C) a flat equation
   D) a population projection table
   Answer: D

Pages: 391
2. Population growth
   A) shows a constant rate of increase every year.
   B) shows a stable age distribution depending on survivorship.
   C) shows a dependency on fecundities in each age class.
   D) all of the above
   Answer: D

Pages: 396
3. Population regulation
   A) does not usually occur in most populations.
   B) results in exponential growth.
   C) is usually density dependent.
   D) occurs when a population is introduced into a new habitat in which it is well suited.
   Answer: C

Pages: 392
4. The rate of increase in population size called r is
   A) directly proportional to instantaneous birthrate minus instantaneous death rate.
   B) is not used in the integrated form.
   C) is not constant during the growth period.
   D) all of the above
   Answer: A

Pages: 393
5. Exponential growth curves
   A) are usually straight lines going upward.
   B) are curved upward.
   C) are usually straight lines going downward.
   D) are mixtures of curved and straight lines.
   Answer: B

Pages: 394

6. Exponential growth cannot continue indefinitely to infinite populations. So populations are limited and the equation which describes this is the
   A) population projection equation.
   B) net reproductive rate.
   C) logistics equation.
   D) intrinsic rate of natural increase.
   Answer: C

Pages: 396

7. The level at which available resources can sustain individuals in a population at survival level is
   A) the mortality factor.
   B) natality.
   C) the carrying capacity.
   D) all of the above
   Answer: C

Pages: 396

8. Some important density dependent factors are
   A) disease.
   B) competition.
   C) immigration.
   D) all of the above
   Answer: D

Pages: 396

9. In scramble competition
   A) all the population survives.
   B) all individuals garner such a small share of the resources none survive.
   C) a contest competition occurs.
   D) occurs only with insects.
   Answer: B

Pages: 400

10. In plant populations, a decline in density results in _____ when resources are scarce.
    A) extinction
    B) exponential growth
    C) self thinning
    D) no changes
    Answer: C

Pages: 402

11. A density independent factor is
    A) weather.
    B) disease.
    C) emigration.
    D) parasites.
    Answer: A

12. Population fluctuations
   A) are absent in all populations examined.
   B) occur yearly or seasonally.
   C) have no reasons to occur.
   D) none of the above
   Answer: B

13. Determination of density dependent influences can be accomplished by
   A) studying resilience.
   B) studying population over long periods of time.
   C) by improving methods of population estimating.
   D) studying key factor analysis.
   Answer: D

14. When <u>R</u> becomes less than one
   A) the population grows arithmetically.
   B) the population reaches the carrying capacity.
   C) the population faces extinction.
   D) the population is stable.
   Answer: C

15. Extinction in wild populations is due to
   A) excessive hunting.
   B) habitat destruction.
   C) stochastic events.
   D) all of the above
   Answer: D

True-False

Write T if the statement is true and F if the statement is false.

1. When death rates are less than birth rates <u>R</u> is positive and growth occurs.
   Answer: True

2. Starvation is a density independent factor.
   Answer: False

3. Exponential growth occurs for a short time when a new species invades a suitable habitat.
   Answer: True

4. The presence of suitable habitat determines the carrying capacity.
   Answer: True

Pages: 393
5. The carrying capacity is the largest number of individuals of a population that can be supported by a habitat.
   Answer: True

Pages: 409
6. Habitat destruction is not a major cause in extinction.
   Answer: False

Pages: 409
7. In small isolated populations, extinction can be a stochastic process.
   Answer: True

Pages: 396
8. A good environment free of predators can support continuous exponential growth.
   Answer: False

Pages: 402
9. Weather is a density-dependent factor.
   Answer: False

Pages: 397
10. When a resource is in short supply competition may act as a density-dependent factor and reduce population growth.
    Answer: True

Pages: 405
11. The oscillations of the snowshoe hare and the Canadian lynx appear to be similar due to predation on the hare by the lynx.
    Answer: True

Short Answer

Write the word or phrase that best completes each statement or answers the question.

Pages: 390
1. The _____ equals the sum for the lifetime of each age specific survivorship times each age specific fecundity.
   Answer: net reproductive rate

Pages: 390
2. If the birthrate equals the death rate the _____ is stable and constant.
   Answer: population

Pages: 391
3. A _____ can be formulated demonstrating population increase.
   Answer: population projection table

Pages: 392
4. In _____ growth, the formula is $\underline{N(t)=N(0)e|rt}$.
   Answer: exponential

Pages: 393
5. Logistics growth occurs when a limiting population number, the _____ occurs.
   Answer: carrying capacity

Pages: 394
6. _____ growth occurs when there is no limiting population or when the limiting population is not being approached and the inflection point has not been reached.
   Answer: Exponential

Pages: 397
7. Density _____ regulation limits a population by intra-specific competition for food, nesting sites, or habitat.
   Answer: dependent

Pages: 397
8. Some density _____ factors are a colder winter than normal and unfavorable environmental conditions.
   Answer: independent

Pages: 404
9. The nature of fluctuation reflects the populations _____.
   Answer: resilience

Pages: 405
10. The Canadian lynx population and the lemming population are two classic examples of _____.
    Answer: cycles

Pages: 406
11. Key factor analysis relates a _____ in the biological or environmental condition associated with mortality that causes major changes in the population.
    Answer: condition or factor   (either answer)

Pages: 407
12. When populations continuously decline _____ may occur.
    Answer: extinction

Matching

Choose the item from Column 2 that best matches each item in Column 1.

Regulation of populations is by the following:

Pages: 402
1. Column 1: Drought in the summer
   Column 2: density-independent

Pages: 396
2. Column 1: Diseases
   Column 2: density-dependent

Pages: 406
3. Column 1: Starvation
   Column 2: density-dependent and
   density-independent

Pages: 396
4. Column 1: Immigration
   Column 2: density-dependent

Pages: 402
5. Column 1: Poisons
   Column 2: density-independent

Pages: 406
6. Column 1: Predators
   Column 2: density-dependent

Pages: 402
7. Column 1: A cold long winter
   Column 2: density-independent

Essay

Write your answer in the space provided or on a separate sheet of paper.

Pages: 392, 393
1. Distinguish between exponential growth and logistics growth using the equations for the basis of your answer.
   Answer: The equation for exponential growth is $\underline{N(t)=N(0)e^{rt}}$ and for logistics growth the equation is $\underline{dN/dt=rN(K-N/K)}$. Logistics growth is a special case of exponential growth which says that $\underline{N}$ approaches $\underline{K}$, the carrying capacity growth becomes zero and the population size is limited at $\underline{K}$. Exponential populations increase at a finite rate.

Pages: 408
2. What factors lead to extinction and how do they operate on growth and growth rate?
   Answer: When $\underline{R}$ becomes less than 1, then the population declines. If declining continues and the species is unable to adapt to new conditions or habitat loss, the population becomes too small for successful reproduction and dies off. Extinctions are deterministic if brought about by some force of change affecting all and stochastic if extinction occurs from random changes in a population.

Pages: 398
3. What do scramble competition, or exploitative competition, and contest competition mean. How are populations affected?
   Answer: Scramble or exploitative competition occurs when each individual is affected by a shortage of a shared resource and individual growth and population growth are reduced. Contest or interference competition occurs when only some of the individuals of a population gain access to the resources.

137

Pages: 406
4. What is key factor analysis and why is it important?

    Answer: A key factor is a biological or environmental condition associated with mortality that causes major fluctuations in population size. Key factor analysis is based on analysis on the summation of $\underline{k}$ values from a life table. The summation of $\underline{k}$ values for all age class is called $\underline{K}$ factor. This information tells what factor in life table is affected.

Pages: 403
5. What fluctuations occur in populations and what causes them?

    Answer: Fluctuations and cycling do not occur in all populations. Fluctuations occur due to lags in density-dependent populations and the resilience of the population. The fluctuations could be due to physiological or environmental factors.

Multiple-Choice

Choose the one alternative that best completes the statement or answers the question.

Pages: 417
1. When additional grain was given to the mourning doves and more space allowed
    A) the dominant male harassed the others.
    B) stress was reduced.
    C) stress was increased as all scrambled for the food.
    D) the codominant female attacked the dominant male.
    Answer: B

Pages: 412
2. Intraspecific competition means competition between
    A) alpha males.
    B) different species.
    C) different populations.
    D) individuals of the same species.
    Answer: D

Pages: 412
3. Plants react to increased density by
    A) decreased growth.
    B) increased growth.
    C) no change in growth.
    D) increased root growth.
    Answer: A

Pages: 413
4. When young birds leave the nest a/an
    A) breeding dispersal occurs.
    B) general dispersal occurs.
    C) natal dispersal occurs.
    D) fetal dispersal occurs.
    Answer: C

Pages: 413
5. Murray's Rule on dispersal is to
    A) go in a straight line.
    B) move to the first uncontested site and no further.
    C) move as far away as possible.
    D) all of the above
    Answer: B

Pages: 414
6. The highest rates of dispersal occur
    A) after peak densities.
    B) at peak densities.
    C) just before peak densities.
    D) at low densities.
    Answer: C

Pages: 415
7. Dispersal advantages include
   A) improved fecundity.
   B) predation is more likely.
   C) unfamiliarity with terrain.
   D) hybrid young not well adapted.
   Answer: A

Pages: 416
8. Saturation dispersal takes place
   A) when disperses are in poor shape.
   B) after carrying capacity has been exceeded.
   C) with mostly juveniles and subdominants.
   D) all of the above
   Answer: D

Pages: 417
9. From studies of wolves populations ecologists concluded that
   A) dispersal regulates populations.
   B) dispersal expanded populations.
   C) dispersal was a rare occurrence.
   D) dispersal involved mostly females.
   Answer: B

Pages: 417
10. Social dominance is based on
    A) intraspecific aggressiveness and intolerance.
    B) mutual attraction.
    C) high tolerance.
    D) positive reaction to crowding.
    Answer: A

Pages: 418
11. Some peck orders are
    A) straight line.
    B) all peck all others.
    C) circular.
    D) all of the above
    Answer: A

Pages: 419
12. Territories are defended by
    A) conflicts.
    B) bird songs.
    C) visual displays.
    D) all of the above
    Answer: D

Pages: 421
13. Phermones
    A) are chemical messengers secreted by endocrine glands.
    B) are sex hormones.
    C) are secreted by the ears of animals.
    D) all of the above
    Answer: A

Pages: 419
14. Territorial size
    A) is usually constant.
    B) usually varies seasonally and yearly.
    C) is inversely related to the size of the animal.
    D) all of the above
    Answer: B

Pages: 423
15. As a result of contest competition among males in many animals, some are denied territory and
    A) this group dies.
    B) this group is composed of debilitated individuals.
    C) this group is a floating reserve.
    D) females control this group.
    Answer: C

Pages: 424
16. The area in which an animal lives
    A) may not be defended.
    B) may not overlap other individuals.
    C) is the home range.
    D) all of the above
    Answer: D

True-False

Write T if the statement is true and F if the statement is false.

Pages: 412
1. Crowded conditions result in stress by increasing social contact.
   Answer: True

Pages: 412
2. Stress results in hormonal changes and changes in the immune system.
   Answer: True

Pages: 412
3. Regulation of plant populations such as low nutrients do not involve stress.
   Answer: False

Pages: 413
4. Stress of crowding is avoided by some animals when dispersal occurs.
   Answer: True

Pages: 413

5. Breeding dispersal occurs when animals are forced into poorer reproductive sites.
   Answer: False

Pages: 416

6. Individuals usually do not disperse until carrying capacity is reached.
   Answer: False

Pages: 424

7. Home ranges have a constant size.
   Answer: False

Pages: 418

8. A territory is a defended area more or less exclusive and maintained by an individual or social group.
   Answer: True

Pages: 416

9. Most populations dispersing at post-saturation (carrying capacity) are density-independent.
   Answer: False

Pages: 419

10. More energy must be expended by an animal as the size of it's territory increases.
    Answer: True

Pages: 420

11. Defending a territory may involve displays, songs, posturing, or combat.
    Answer: True

Pages: 417

12. Social hierarchies increase the aggressive behavior of a population by establishing a peck order.
    Answer: False

Pages: 418

13. The male leader in a pack of wolves is call the alpha male.
    Answer: True

Short Answer

Write the word or phrase that best completes each statement or answers the question.

Pages: 412

1. Stress in vertebrates can act on the individual by a/an _____ feedback involving the functioning of the _____ and _____ glands.
   Answer: physiological, pituitary, adrenal   (last two any order)

Pages: 412

2. Increasing population _____ stress and profound hormonal changes occur suppressing _____ and _____.
   Answer: density, growth, reproduction   (last two any order)

3. High population density can result in _____ births and increased _____ .
Answer: decreased, mortality (in order)

4. Some plants such as ragweed and other ruderals respond to stress by slowing _____ and production of _____ .
Answer: growth, seeds (in order)

5. _____ is the permanent movement an individual makes from its birth site to the place where it reproduces or would reproduce if it survived.
Answer: Dispersal

6. Two types of dispersal are _____ and _____ .
Answer: natal, breeding (any order)

7. A motivation to disperse might be _____ competition.
Answer: intraspecific

8. _____ occurs when the carrying capacity has been exceeded.
Answer: Saturation dispersal

9. In social dominance, there are two opposing forces: _____ and _____ .
Answer: mutual attraction, social intolerance (any order)

10. Social dominance results in a/an _____ .
Answer: peck order

11. A/an _____ is a defended area maintained by an individual or a group of animals.
Answer: territory

12. _____ is an area where an animal normally lives.
Answer: Home range

Matching

Choose the item from Column 2 that best matches each item in Column 1.

1. Column 1: Stress which is _____ can act on physiological feedback.
Column 2: density-dependent

2. Column 1: Dispersal during the reproductive period is due to a _____ mechanism.
Column 2: density-independent

Pages: 414
3. Column 1: Christianson hypothesized that in rodents aggression was _____.
   Column 2: density-dependent

Pages: 416
4. Column 1: Dispersal after carrying capacity is reached is _____.
   Column 2: density-dependent

Pages: 417
5. Column 1: Intraspecific aggressiveness and individual intolerance can result in _____.
   Column 2: social dominance

Essay

Write your answer in the space provided or on a separate sheet of paper.

Pages: 412,417
1. How does density affect population growth and behavior?
   Answer: As populations grow crowding occurs introducing stress that results in hormonal changes and changes in the immune system which can result in increased mortality and reduced natality and increased social interactions. Although dispersal is not a regulatory mechanism it can expand the population.

Pages: 413,414
2. What causes dispersal?
   Answer: Dispersal is innate to populations. Young animals undertake a natal dispersal away from birth place, seeking new habitat. Adults may disperse to seek better reproductive sites. Some dispersal may be in response to overcrowding or lack of food.

Pages: 417-419
3. Distinguish between territorial and social animals.
   Answer: Social dominance develops out of aggressive interactions. It results in territorial formation of a hierarchy or peck order. Animals establish and defend an area of space. Those not having a territory or those not in the higher positions of the heirarchy are excluded from the resources.

Pages: 419,420
4. What behaviors are involved in territoriality?
   Answer: Maintenance of a territory requires a great expenditure of energy, the larger the territory the greater the energy. Territories can be defended by attacking the intruder (birds and other animals), aggressive displays (prairie chickens and most animals), vocalizations (frogs, birds), and visual displays (wolves, deer, birds).

CHAPTER 20 Life History Patterns

Multiple-Choice

Choose the one alternative that best completes the statement or answers the question.

Pages: 428
1. A mating system does not include
   A) the number of mated males and females acquired.
   B) the manner in which mates are acquired.
   C) asexual reproduction.
   D) pair bonding.
   Answer: C

Pages: 428
2. Dioecious individuals
   A) separate sexes on different organisms.
   B) have both sexes on the same organism.
   C) have asexual reproduction.
   D) all of the above
   Answer: A

Pages: 428
3. Monoecious individuals
   A) separate sexes on different organisms.
   B) have both sexes on the same organisms.
   C) have asexual reproduction.
   D) none of the above
   Answer: B

Pages: 428
4. Which of the following is a hermaphrodite:
   A) snake
   B) lizard
   C) earthworm
   D) dogs
   Answer: C

Pages: 428
5. In polyandry
   A) males control access to females directly.
   B) neither sex has opportunity of monopolizing additional members of the opposite sex.
   C) females control access to males.
   D) there are explosive breeding assemblages.
   E) males aggregate on leks.
   Answer: C

Pages: 430
6. Polygamy
   A) means acquiring two or more mates.
   B) includes polygyny.
   C) includes polyandry.
   D) all of the above
   Answer: D

Pages: 431
7. Sexual selection
   A) may not be related to fitness.
   B) is exclusively a male choice.
   C) is probably related to natural selection.
   D) is not a function of natural selection.
   Answer: C

Pages: 431
8. The handicap hypothesis postulates among other factors is a
   A) female handicap.
   B) male handicap.
   C) general inviability trait.
   D) male preference.
   Answer: B

Pages: 432
9. Intrasexual selection results in
   A) male to male competition.
   B) male to female competition.
   C) results in elaborate plumage in females.
   D) all of the above
   Answer: A

10. On the Galapagos Islands, female cactus finches select a mate
    A) by the black plumage of the male.
    B) on the basis of courtship performance.
    C) in such a way that they mate with experienced males.
    D) all of the above
    Answer: D

Pages: 433
11. Among polygamous birds
    A) there are no strict mate preferences.
    B) most do not mate at all.
    C) selection is based on territory size and quality.
    D) all of the above
    Answer: C

Pages: 433
12. The long-tailed widowbird of the African savanna was studied by the ecologist Anderson who with judicious clipping and glueing found
    A) the longer the tail the poorer the reproductive success.
    B) long tailed birds could not defend the territory.
    C) remarkable enough females selected males with clipped tails most frequently.
    D) the longer the tails the more mates were attracted and reproductive success was higher.
    Answer: D

Pages: 435
13. Which of the following has not been advanced as a theory for lek behavior?
    A) female choice
    B) hotspot model
    C) hotshot model
    D) male choice
    Answer: D

Pages: 437
14. When the reproductive effort of an individual is increased there is a _____ in survivorship.
    A) increase
    B) constant survivorship
    C) decrease
    D) fluctuation
    Answer: C

Pages: 438
15. For iteroparous organisms early reproduction means
    A) earlier maturity.
    B) reduced survivorship.
    C) less growth.
    D) all of the above
    Answer: D

Pages: 441
16. For organisms that produce many young
    A) little parental care is taken of young.
    B) males care for young.
    C) females care for young.
    D) all of the above
    Answer: A

Pages: 444
17. Many hermaphroditic plants
    A) are self-sterile.
    B) are self-fertile but rarely self-pollinated.
    C) are self-fertile and are self-pollinated.
    D) all of the above
    Answer: D

18. A population characterized by delayed reproduction, slower growth rate, and large body size, and that is efficient, is
    A) r-selected.
    B) K-selected.
    C) are semelparous.
    D) a lek.
    Answer: B

19. When the females in a population produce one group of progeny early the population is
    A) r-selected.
    B) K-selected.
    C) unselected.
    D) q-selected.
    Answer: A

20. The production of large numbers of young is found in
    A) elephants.
    B) perennial plants.
    C) K-selected populations.
    D) annual plants.
    Answer: D

True-False

Write T if the statement is true and F if the statement is false.

1. Plants with separate male and female individuals are called monoecious.
   Answer: False

2. Hermaphroditic plants are said to be monoecious.
   Answer: True

3. Sexual reproduction involves formation of haploid gametes, eggs, and sperm that combine to form a dipoid cell or zygote thus recombining the genes and increasing variability.
   Answer: True

4. Polygyny is when individual females control or gain access to individual males.
   Answer: False

5. Polyandry is when individual males control or gain access to individual females.
   Answer: False

6. Monogamy is the formation of a pair bond between one male and one female.
   Answer: True

Pages: 430

7. Intrasexual competition among males results in the males having choice of the females.
   Answer: False

Pages: 430

8. Among birds, females have a strong preference for males with high-quality territories.
   Answer: True

Pages: 433

9. In some polygamous species, the females mate only with the highest ranking males.
   Answer: True

10. In many species of birds male fitness and hence favorable female selection is based on a dull and very plain camouflage plumage.
    Answer: False

Pages: 436

11. Lek species gather together and the females defend large areas.
    Answer: False

Pages: 437

12. Reproduction can shorten life expectancy.
    Answer: True

Pages: 438

13. Semelparous organisms reproduce constantly throughout their life.
    Answer: False

Pages: 441

14. Reproducing organisms have two choices to apportion parental investment: have many small young or a few large ones.
    Answer: True

Pages: 448

15. Species are r-strategists if they are long lived and have a late, low reproductive rate.
    Answer: False

Short Answer

Write the word or phrase that best completes each statement or answers the question.

Pages: 428

1. _____ plants have separate male and female plants, but _____ (hermaphroditic) species with both sexes on the same plant occur.
   Answer: Dioecious, monoecious   (in order)

Pages: 428

2. _____ refers to the number of mates a male or female acquires, the manner in which they are acquired, the nature of the pair bond, and the pattern of parental care by each mate.
   Answer: Mating system

Pages: 430

3. In _____ , the formation of a pair bond is between one male and one female.
   Answer: monogamy

Pages: 430

4. In _____ an individual female gains control of or access to two or more males.
   Answer: polyandry

Pages: 430

5. In _____ an individual male gains access to or control of one or more females.
   Answer: polygyny

Pages: 430

6. A special form of _____ occurs in promiscuity in which males or females copulate with many
   of the opposite sex and form no pair bonds.
   Answer: polygamy

Pages: 430

7. In most species there is an intense rivalry between males, _____, for a mate but the female
   usually does the _____.
   Answer: intrasexual competition, sexual selection   (in order)

Pages: 432

8. For _____ females the selection is based on the territory held by the male.
   Answer: monogamous

Pages: 435

9. Color, plumage, past success, and behavior in leks are all mechanisms of _____.
   Answer: sexual selection

Pages: 436

10. _____ is the amount of time and energy an organism expends in reproduction.
    Answer: Reproductive effort

Pages: 437

11. For polygamous males, defending females, fighting off male competitors, and sexual activity
    shorten _____ and future _____.
    Answer: life expectancy, reproductive activity   (in order)

Pages: 437

12. Species producing millions of progeny rarely have _____.
    Answer: parental care

Pages: 450

13. Reproductive success depends on the choice of a mate and the choice of _____.
    Answer: habitat

Pages: 448

14. With stable populations of long lived individuals such as forests _____ is used for
    reproduction.
    Answer: k-strategy

15. Populations in areas of high mortality and unpredictable environments have a/an _____ for reproduction.
   Answer: r-strategy

Matching

Choose the item from Column 2 that best matches each item in Column 1.

Species can be either r-selective or K-selective and have different selection pressures on them.

Pages: 448
1. Column 1: Fast growing
   Column 2: r-selective

Pages: 448
2. Column 1: Reproduce each season
   Column 2: K-selective

Pages: 448
3. Column 1: Fast reproduction rates
   Column 2: r-selective

Pages: 448
4. Column 1: Long lived
   Column 2: K-selective

Pages: 448
5. Column 1: Remain around carrying capacity
   Column 2: K-selective

Pages: 448
6. Column 1: Are found in early successional areas
   Column 2: r-selective

Essay

Write your answer in the space provided or on a separate sheet of paper.

Pages: 430-433
1. Monogamy is common in birds along with polygyny which is also prevalent. The situation is reversed in mammals. Why are these two mating systems prevalent in birds and mammals? In your answer, include the adaptive significance of each.
   Answer: Monogamy occurs in birds because the male is necessary to feed and protect the young birds thus adaptive behavior includes mamles feding young by not all birds are dependent on food from both parents in species feeding on grain polygyny occurs. In mammals, polygyny is common because the mammary glands feed the young after birth and the male is not needed.

Pages: 432-434
2. Birds have unusual plumage (tails and special feathering) and courtship rituals before copulation occurs. What importance are these in terms of natural selection and adaptation?
   Answer: The importance of these displays is that they are adaptive showing increased fitness of males. How the increased fitness is related is not always known.

Pages: 435
3. What is lek behavior? How has it come about?
   Answer: Lek behavior occurs because the individuals assemble on a communal courtship ground and a small mating territory is defended by males with females choosing mates based on a series of displays. There are three hypotheses which explain lek behavior: female choice of place to mate, hotspot model males cluster where females are high, and hotshot model strong hierarchies.

Pages: 428,444
4. When many plants reproduce asexually why is sexual reproduction important in plants?
   Answer: Sexual reproduction is important in plants because it maintains genetic variability in a population necessary to adapt to environmental changes through time.

Pages: 446
5. Why are sex ratios skewed toward one sex or the other in many populations?
   Answer: There are several reasons: Differential mortality between sexes; males have a higher mortality rate. Nourishment and environmental conditions may favor the ratio of one sex over another at birth, especially among polygamous animals. Temperature can influence the sex at hatching of reptiles. Lower incubation temperatures favors males.

Multiple-Choice

Choose the one alternative that best completes the statement or answers the question.

Pages: 458
1. Gametes are formed after
   A) mitosis.
   B) fertilization.
   C) mutation.
   D) meiosis.
   Answer: D

Pages: 458
2. Mutations occur
   A) as a result of selection.
   B) as a result of surroundings.
   C) to enable population expansion.
   D) by chance.
   Answer: D

Pages: 457
3. The genetic constitution of an individual is its
   A) phenotype.
   B) genotype.
   C) diploid.
   D) mutation.
   Answer: B

Pages: 459
4. In a Hardy-Weinberg population the fraction of recessive phenotypes is 0.01, therefore the frequency of dominant phenotypes in the population is _____.
   A) 0.005
   B) 0.025
   C) 0.99
   D) 0.10
   Answer: C

Pages: 465
5. Which of the following is not an assumption of the Hardy-Weinberg equilibrium.
   A) no random mating
   B) no mutation
   C) no immigration
   D) no natural selection
   Answer: A

Pages: 458
6. Variation in populations due to inheritance is
   A) chromosome rearrangement.
   B) polyploidy.
   C) mutation.
   D) all of the above
   Answer: D

Pages: 458
7. Variation in phenotype due to factors which are not genetic can be due to
   A) altruistic genes.
   B) Hardy-Weinberg treatments.
   C) environment.
   D) all of the above
   Answer: C

Pages: 466
8. In some birds (scrub jays) and mammals (black-backed jackal) the young stay with the parents an extra year and assist in rearing the newborns. This behavior is
   A) not beneficial to the adults.
   B) beneficial to the young.
   C) altruistic and increases all individuals fitness
   D) all of the above
   Answer: C

Pages: 461
9. Natural selection
   A) maintains genetic variation.
   B) those that are not as fit, do not transfer their genes to future generations and are not involved in natural selection.
   C) is basically non-random reproduction.
   D) does not occur in insects.
   Answer: C

Pages: 462
10. Selection may be
    A) disruptive.
    B) directional.
    C) stabilizing.
    D) all of the above
    Answer: D

Pages: 463
11. Inbreeding
    A) is the result of mating between unrelated individuals but are of the same species.
    B) decreases homozygosity.
    C) results in undesirable phenotypes due to increased appearance of recessive genotypes.
    D) all of the above
    Answer: C

Pages: 470
12. In normally outcrossing populations
    A) homozygosity increases.
    B) rare deleterious genes become expressed.
    C) close inbreeding is detrimental.
    D) all of the above
    Answer: C

Pages: 472
13. Effective population size refers to
    A) all the organisms in the population.
    B) a deme of organisms.
    C) the males in the population.
    D) the number of adults contributing gametes to the next generation.
    Answer: D

Pages: 473
14. A population facing over exploitation experiences a/an _____ when the only survivors have a sample of the total gene pool.
    A) bottleneck
    B) sampling experience
    C) extinction
    D) increase in gene pool
    Answer: A

Pages: 473
15. In populations covering a large area, the effective breeding size is
    A) very large.
    B) very small.
    C) neighborhood size.
    D) does not exist, they breed all over.
    Answer: C

Pages: 474
16. Founder effect
    A) occurs due to selection.
    B) is produced when very large populations migrate and start new populations.
    C) is produced by random genetic drift.
    D) all of the above
    Answer: C

Pages: 475
17. Which of the following statements is true.
    A) Monogamous populations retain rare alleles.
    B) Cycles of inbreeding and outbreeding retain alleles.
    C) Effective population size is more important in exploited populations.
    D) all of the above
    Answer: D

Pages: 458
18. Polyploidy
    A) is frequent in humans.
    B) is frequent in animals.
    C) occurs mostly in plants.
    D) all of the above
    Answer: C

Pages: 455
19. Factors producing genetic drift are
    A) effective population size is small.
    B) gene flow.
    C) mating strategies.
    D) all of the above
    Answer: D

True-False

Write T if the statement is true and F if the statement is false.

Pages: 456
1. An individual of genotype <u>AA</u> is said to be homozygous.
   Answer: True

Pages: 456
2. Variations in a specific chararcter which separate a population into discrete classes are continuous variations.
   Answer: False

Pages: 456
3. The genetic information an individual has is its phenotype.
   Answer: False

Pages: 457
4. Genes of a pair that control a character in a different way are called alleles.
   Answer: True

Pages: 458
5. When the genotype is <u>Aa</u>, the organisms is said to be heterozygous.
   Answer: True

Pages: 458
6. The diploid number refers to the number of chromosomes in a gamete.
   Answer: False

Pages: 459
7. Loss of a portion of a chromosome is called a deletion.
   Answer: True

Pages: 458
8. Polyploidy is a condition of having extra sets of chromosomes.
   Answer: False

Pages: 460

9. The Hardy-Weinberg equilibrium occurs when there are no mutations, no immigrations or emigration, and no selective mating.
Answer: True

Pages: 460

10. In the Hardy-Weinberg equilibrium, the allele and genotype frequencies remain the same from generation to generation.
Answer: False

Pages: 458

11. Mutations are changes in the genetic code which are inheritable.
Answer: True

Pages: 461

12. DDT was a selective agent in selecting for DDT resistance.
Answer: True

Pages: 461

13. If an organism survives only as an individual and does not leave mature reproducing progeny. It is said to be adapted because it survived.
Answer: False

Pages: 462

14. Stabilizing selection favors one or the other extreme phenotypes.
Answer: False

Pages: 468

15. Kin selection involves inbreeding.
Answer: True

Short Answer

Write the word or phrase that best completes each statement or answers the question.

Pages: 456

1. The most obvious variations in a population are _____ variations which separated individuals into discrete categories such as male and female.
Answer: discontinuous

Pages: 456

2. Another type of variation is the _____ variation which places individuals within a range of values such as shapes and sizes of petals and sepals.
Answer: continuous

Pages: 456

3. The _____ is the sum of the genetic information carried by an individual or more simply its gene formula.
Answer: genotype

4. The external visible expression of genes in the environment is the _____ or what an organism looks like with regard to its genotype.
Answer: phenotype

5. Organisms give rise to a range of phenotype expression in different environments known as

_____.

Answer: phenotype plasticity

6. Normal body cells contain the _____ number of chromosomes but as a result of a special division called _____ which takes place in reproductive tissue, cells with one set of chromosomes characteristic of the species, called _____ cells are produced.
Answer: diploid, meiosis, haploid (in order)

7. Each gene occupies a locus on the haploid set of chromosomes and when both chromosomes present in the adult carry two genes the same the individual is _____.
Answer: homozygous

8. If the genes of a gene pair differ the individual is called _____.
Answer: heterozygous

9. Sources of genetic variation and phenotypic variation in a population are _____(name all).
Answer: gene mutation, new combinations, chromosome aberrations, polyploidy (any order)

10. The _____ equilibrium states that the genotype frequency and allele frequency in a population remain the same from generation to generation if nothing occurs to alter genetic composition of the population.
Answer: Hardy-Weinberg

11. In house flies, the resistance to DDT became common as treatments with DDT continued in a phenomena called _____.
Answer: natural selection

12. The _____ of an individual measured by it's reproducing offspring.
Answer: fitness

13. _____ occurs when an individual's fitness decreases but fitness of close relatives increases due to an altruistic trait.
Answer: Kin selection

14. The _____ measures the probability of having alleles identical by descent from common ancestors as in brother-sister matings or other relative matings.
Answer: inbreeding coeficient

Pages: 471
15. _____ is the chance fluctuation in allele frequencies in small populations as a result of random sampling.
    Answer: Genetic drift

Pages: 475
16. The threshold number of individuals necessary to keep a viable population is the _____.
    Answer: minimum viable population

Matching

Choose the item from Column 2 that best matches each item in Column 1.

1. Column 1: Favors both extremes of the phenotype
   Column 2: disruptive selection

2. Column 1: Favors average phenotype
   Column 2: stabilizing selection

3. Column 1: Favors one extreme phenotype
   Column 2: directional selection

4. Column 1: Development of polymorphic species in the pepper moth of England.
   Column 2: disruptive selection

Essay

Write your answer in the space provided or on a separate sheet of paper.

Pages: 456
1. Distinguish between genotype and phenotype? How does natrunal selection work with the genotype and phenotype?
   Answer: The genotype refers to the genes possessed by an individual. The phenotype is the external expression of that genotype. Natural selection works on the phenotype. By favoring certain phenotypes over other natural selection influence allele and genotypic frequency.

Pages: 457
2. What are the major sources of variation in the gene pool?
   Answer: The major sources of variation in the gene pool are gene or point mutations, chromosomal abberrations, and polyploidy. These involve permanent change and random rearrangement of existing genes by recombination and also produces variation.

Pages: 458
3. What is the Hardy-Weinberg law and why is it important?
   Answer: The Hardy-Weinberg law is a statement that says there is no gene or allele frequency change over many generations as long as random mating, no mutations, and no emigration or immigration. In other words, some change must occur before genes change in a population.

Pages: 461
4. What is meant by fitness?

   Answer: Fitness is measured by the number of reproducing offspring. Individuals that contribute the most to the gene pool are said to be the most fit.

Pages: 467
5. What is inbreeding and what does it do?

   Answer: Inbreeding is mating with relatives. It can result in inbreeding depression as measured by expression of deleterious genes, poor growth, decreased fertility, decreased fecundity, small body growth, and premature death.

Multiple-Choice

Choose the one alternative that best completes the statement or answers the question.

Pages: 480
1. Interspecific competition is an interaction between two species which is
   A) beneficial to both.
   B) beneficial to one but not affecting the other.
   C) beneficial to one but negative to the other.
   D) detrimental to both.
   Answer: D

   + is the sign for a favorable reaction, - is the sign for a detrimental reaction, and O is the sign for no reaction. Answer the following problems:

Pages: 480
2. Parasitism is
   A) ++
   B) +-
   C) --
   D) +O
   Answer: B

Pages: 480
3. Mutualism is
   A) ++
   B) +-
   C) --
   D) +O
   Answer: A

Pages: 480
4. Commensalism is
   A) ++
   B) OO
   C) --
   D) +O
   Answer: D

Pages: 480
5. Neutralism is
   A) ++
   B) --
   C) OO
   D) O+
   Answer: C

Pages: 480
6. The two kinds of competition are
   A) amensalism and commensalism.
   B) amensalism and neutralism.
   C) amensalism and predation.
   D) exploitative and interference.
   Answer: D

Pages: 480
7. According to the Lotka and Volterra competition equation
   A) competing species grow faster and reach higher carrying capacity when grown together.
   B) one grows well and the other does not.
   C) both grow when together but not as fast and neither reaches carrying capacities as large as
      when they were grown separately.
   D) B and C
   Answer: D

Pages: 483
8. Another useful plotting technique devised by Lotka and Volterra was to plot
   A) $\underline{N}(1)$ vs $\underline{N}(2)$.
   B) $\underline{N}$ vs $\underline{K}$ for each population.
   C) $\underline{K}$ vs $\underline{K}$ for each population.
   D) $\underline{R}$ vs $\underline{R}$ for each population.
   Answer: A

Pages: 484
9. _____ tested Lotka-Volterra equations in the laboratory using paramecia.
   A) R.L. Smith
   B) Lotka
   C) Volterra
   D) Gause
   Answer: D

Pages: 487
10. The competitive exclusion principle states that _____ using the same resource.
    A) two competing populations can coexist
    B) no two competing species can coexist
    C) two competing populations can almost exist
    D) all of the above
    Answer: B

Pages: 495
11. When two organisms use only a part of a resource at the same time a condition of _____
    occurs.
    A) niche overlap
    B) mutualism
    C) amensalism
    D) commensalism
    Answer: A

Pages: 492
12. The niche that an organism has is
    A) where it is found.
    B) where it finds food.
    C) where it lives.
    D) all of the above
    Answer: D

Pages: 495
13. The realized niche is
    A) a demonstrable reality.
    B) an abstraction.
    C) is confined to feeding.
    D) all of the above
    Answer: B

Pages: 495
14. Putwain and Harper studied two species of dock, Rumex acetosa and Rumex acetosella by treating the flora with specific herbicides. They found
    A) R. acetosa spread rapidly after grasses were removed.
    B) R. acetocella spread only after grasses and other forbs were eliminated.
    C) Both had niche overlap.
    D) all of the above
    Answer: D

Pages: 498
15. Some characteristics of niches are
    A) niche widths.
    B) niche compressions.
    C) niche shifts.
    D) all of the above
    Answer: D

Pages: 498
16. Allelopathy is a form of interference competition in which _____ use chemicals to inhibit other species.
    A) fish
    B) invertebrates
    C) plants
    D) all of the above
    Answer: C

Pages: 480
17. Amensalism
    A) is a form of competition.
    B) is a condition of two populations when one is negatively affected and the other is unaffected.
    C) is limited to plants.
    D) all of the above
    Answer: B

True-False

Write T if the statement is true and F if the statement is false.

Pages: 480
1. Mutualism is an interaction which is beneficial to both species.
   Answer: True

Pages: 480
2. Epiphytes growing on trees but not harming the trees are commensals.
   Answer: False

Pages: 480
3. Interaction between species is intraspecific competition.
   Answer: False

Pages: 480
4. Interspecific competition may be in the form of an interference rather than exploitative.
   Answer: True

Pages: 482
5. Gause's principle states that two species with identical requirements can occupy the same environment.
   Answer: False

Pages: 482
6. There can be directional interactions favoring one species as predicted by Lotka and Volterra.
   Answer: True

Pages: 482
7. In the Lotka-Volterra model, there are conditions in which neither species can exclude the other in two species competition.
   Answer: True

Pages: 487
8. Classical competition assumes that the environments and competition are variable.
   Answer: False

Pages: 487
9. Competitive exclusion is easy to demonstrate due to interactions with the environment.
   Answer: False

Pages: 488
10. Biologists have concluded that by competitive exclusion, some mallards from more productive wetlands have contributed to the decline of the black duck in southern Ontario.
    Answer: True

Pages: 491
11. Resource partitioning is very uncommon in nature, particularly in great tits and coal tits.
    Answer: False

12. According to Whittaker, Lavine, and Root, the niche and the habitat compose the ecotope, how a species conforms to its environment.
    Answer: True

13. Niche overlap is very uncommon in nature.
    Answer: False

14. Observations of a number of species sharing the same habitat suggest that they coexist by utilizing different resources.
    Answer: True

15. The sum of all resources exploited by an organism is the niche width.
    Answer: True

Short Answer

Write the word or phrase that best completes each statement or answers the question.

1. Among the two species population interaction _____ is when one species kill and eats another and a +- designation is used.
   Answer: predation

2. Positive interactions (++) benefiting both populations are called _____.
   Answer: mutualism

3. _____ (+-) harms host, but is necessary for survival for the other.
   Answer: Parasitism

4. _____ (--) is of no benefit to either population and is actually detrimental to both.
   Answer: Competition

5. The _____ model describes the growth of one species in the presence of another competing species.
   Answer: Lotka-Volterra

6. _____ principle states that two species with the same requirements cannot occupy the same environment.
   Answer: Gause's

7. The general statement that no two species with identical ecological requirements cannot occupy the same area is the _____ principle.
   Answer: competitive exclusion

Pages: 485
8. In two species of paramecium, _____ competition results in the loss of one species.
   Answer: interspecific

Pages: 489
9. _____ refers to the interspecific competition with the production and release of chemical substances.
   Answer: Allelopathy

Pages: 489
10. _____ is an outcome of interspecific competition between several species when differential resource utilization occurs.
    Answer: Resource partitioning

Pages: 494
11. A _____ is an organism's place and function in the environment.
    Answer: niche

Pages: 496
12. Some niche space is shared and some is not when _____ occurs.
    Answer: niche overlap

Pages: 494
13. The _____ is the environment occupied by a species.
    Answer: habitat

Pages: 482
14. If two competing species occupy the same niche, a possible outcome might be _____ or a neutral interaction.
    Answer: coexistence

Pages: 482
15. In this type of competitive relationship the minority component is at an advantage: This interaction is called _____.
    Answer: stabilizing interaction

Matching

Choose the item from Column 2 that best matches each item in Column 1.

Pages: 480
1. Column 1: Bees and birds pollinate flowers.
   Column 2: mutualism

Pages: 480
2. Column 1: Walnut trees produce toxin which kills other plants.
   Column 2: amensalism

Pages: 480
3. Column 1: A tick sucks human blood.
   Column 2: parasitism

Pages: 480
4. Column 1: A hawk kills and eats a mouse.
   Column 2: predation

Pages: 480
5. Column 1: A squirrel stores nuts in the holes of trees.
   Column 2: commensalism

Essay

Write your answer in the space provided or on a separate sheet of paper.

Pages: 480-483
1. Discuss Lotka and Volterra equations and the possible outcomes.
   Answer: The Lotka-Volterra competition model is based on the logistic equations paired with
   competitive effects factored in. The possible outcomes are: 1-Neutral interaction, a
   balance of the two species, 2-Directional interacting in favor of one species,
   3-Stabalizing with the minority at an advantage, 4-Disruptive, majority at an
   advantage.

Pages: 482
2. What is the competitive exclusion principle? How could it be demonstrated?
   Answer: Two species with the same ecological requirements cannot occupy the same environment.
   By studying animals utilizing the same resource over time, best demonstrated in lab
   studies.

Pages: 494
3. What is the niche in theory and in practice?
   Answer: In theory, the niche is the functional role in the community. In theory the
   fundamental niche includes the full range of variables defining the niche and the
   realized niche is the conditions under which the organism actually lives.

Pages: 489
4. What is resource partitioning and why is it important?
   Answer: Resource partitioning is when each species exploits a portion of the resources which
   then become unavailable to the other species. This leads to a practical realization
   that a niche is a multidimensional subject and two competing species do not compete
   for all of the same resources.

Pages: 489
5. What is allelopathy? Is this the ultimate competitive exclusion?
   Answer: Allelopathy occurs in plants where one species produces a toxin that kills other
   species. If the other species are competitors then it is effective, but the toxin
   also kills non-competitors.

CHAPTER 23 Predation

Multiple-Choice

Choose the one alternative that best completes the statement or answers the question.

Pages: 502
1. In a graph of population vs. time in the Lotka-Volterra equations
   A) two straight lines are produced.
   B) one straight line and one oscillating line are produced.
   C) two in phase oscillating lines are produced.
   D) two oscillating lines nearly 90 degrees out of phase are produced.
   Answer: D

Pages: 502
2. The Lotka-Volterra equations are based on the following fact/facts:
   A) when prey populations increase, predator population increase.
   B) when predator populations increase, prey populations decrease.
   C) when prey populations decrease, predator populations increase.
   D) all of the above
   Answer: D

Pages: 503
3. Gause experimented with culture predation in paramecia and found
   A) predators increase as prey decrease.
   B) predators decrease as prey increase.
   C) irregardless of the start, the predator exterminates the prey.
   D) all of the above
   Answer: C

Pages: 506
4. Predator prey studies
   A) new prey or hiding places for prey must be supplied if the populations are to be sustained.
   B) predators can increase even when prey are scarce.
   C) prey decrease even when predators are scarce.
   D) all of the above
   Answer: A

Pages: 508
5. A Type I functional response is
   A) the number of prey taken increases as prey density decreases.
   B) as prey density increases the number of prey taken increases.
   C) as prey number increases, prey taken increases.
   D) all of the above
   Answer: B

Pages: 506
6. In studies of mites, Huffaker found
   A) predators cannot survive when prey population is low.
   B) a self-sustained predator prey population cannot be sustained without emigration.
   C) intensity of predation did not influence oscillation period but prey recovery did.
   D) all of the above
   Answer: D

Pages: 509

7. The Type II response
   A) is identical to the Michaelis-Menten equation.
   B) is found in many invertbrates.
   C) the number of prey taken rises at a diminishing rate.
   D) all of the above
   Answer: D

Pages: 510

8. The Type III response:
   A) The number of prey taken per predator increases with increasing density of prey and levels off.
   B) The number of prey taken per predator decreases with density.
   C) The number of prey taken decreases.
   D) none of the above
   Answer: A

Pages: 510

9. The aggregative response occurs
   A) when prey cluster together.
   B) when prey emigrate in groups.
   C) when predators cluster in areas of high prey density.
   D) all of the above
   Answer: C

Pages: 513

10. In addition to a functional response predators might exhibit
    A) emigration.
    B) immigration.
    C) an increase or decrease in rate of natality or mortality.
    D) all of the above
    Answer: D

Pages: 513

11. In a long term study of kestrels, an increase in the vole population in one region stimulated the raptors to immigrate to areas of high vole population. This is a
    A) numerical response.
    B) Type I functional response.
    C) Type II functional response.
    D) Type III functional response.
    Answer: A

Pages: 514

12. Foraging in a rich food patch and leaving when it is no longer profitable is
    A) catching and killing prey in low prey areas.
    B) optimal foraging strategy.
    C) minimal foraging strategy.
    D) all of the above
    Answer: B

Pages: 515

13. In addition to other considerations most predators have preferred foods and try to achieve
   A) a multiple diet.
   B) an optimal diet.
   C) a minimal diet.
   D) a state of ignoring the diet.
   Answer: B

Pages: 515

14. The length of time a forager should profitably stay at a resource patch before it seeks another, according to the marginal value theorem
   A) relates to richness of the patch.
   B) is time required to extract the resource.
   C) is time required to get there.
   D) all of the above
   Answer: D

Pages: 512

15. Switching prey
   A) is involved in Type III response.
   B) is involved in changing predator choice to a less abundant prey.
   C) the predator pays most attention to rare prey.
   D) is involved in Type I response.
   Answer: A

Pages: 512

16. Tinbergen studied woodland birds and insect prey. He found that when a new prey first appears in an area its risk of being prey is low because
   A) birds are interested in other prey.
   B) birds are not interested in prey.
   C) birds do not have a search image.
   D) birds are lazy.
   Answer: C

Pages: 519

17. Predators follow
   A) risk-sensitive foraging.
   B) the expected energy budget rule.
   C) predation risk foraging.
   D) all of the above
   Answer: D

True-False

Write T if the statement is true and F if the statement is false.

Pages: 502

1. The Lotka-Volterra equations is based on the assumption that the growth rate of the predator was independent of the density of the prey.
   Answer: False

Pages: 502
2. The Lotka-Volterra equation produces coupled oscillations in predation and prey populations which are out of phase.
Answer: True

Pages: 503
3. Gause showed that the predators would over exploit the prey, then exterminate the prey and finally die themselves.
Answer: True

Pages: 503
4. Gause demonstrated that the data on paramecium was consistent with the Lotka-Volterre model.
Answer: False

Pages: 503
5. Gause was able to demonstrate that for sustained predator prey populations, prey must be constantly added.
Answer: True

Pages: 503
6. The recovery of a prey population is dependent on a high population of predators.
Answer: False

Pages: 508
7. The functional response depends on the predator taking more prey when prey density increases.
Answer: True

Pages: 508
8. Type I response involves density-dependent mortality of prey.
Answer: False

Pages: 508
9. In Type II response the number of prey taken rises at a decreasing rate till a maximum is reached.
Answer: True

Pages: 510
10. Type III response can stabilize some prey populations.
Answer: True

Pages: 510
11. In the aggregative response, predators tend to aggregate in areas of low prey density out of habit.
Answer: False

Pages: 510
12. In the numerical response, predators leave when prey density increases.
Answer: False

Pages: 514
13. In order for an animal to obtain a sufficient environment an optimal foraging strategy is used.
Answer: False

Pages: 515
14. The marginal value theorem is used to measure a patch.
Answer: False

Pages: 519
15. Predation risk is the risk of an accident when chasing prey.
Answer: False

Short Answer

Write the word or phrase that best completes each statement or answers the question.

Pages: 502
1. When a rabbit eats grass it is called _____.
Answer: herbivory

Pages: 502
2. When one animal kills and eats another it is called _____.
Answer: predation

Pages: 502
3. The Lotka-Volterra model of predation is a combination of two paired equations, one each for _____ and _____.
Answer: predator, prey (any order)

Pages: 503
4. The ecologist A.J. Nicholson and the mathematician W. Bailey developed a model for _____ relationship.
Answer: host-parasitoid

Pages: 503
5. Rosenweig and McArthur developed a series of _____ to describe the relationship.
Answer: graphical models

Pages: 503
6. The Rosenweig and McArthur model can produce _____ and _____ if the predator and prey curves are altered.
Answer: damped, unstable cycles (any order)

Pages: 505
7. G.F. Gause found that in paramecium and a predatory didinium that the predator always _____ the prey.
Answer: exterminated

Pages: 502
8. When the predator and the prey were the same species _____ occurs.
Answer: cannibalism

Pages: 508
9. When a/an _____ occurs each predator may have taken more prey or taken them sooner.
Answer: functional response

Pages: 508

10. A/an _____ occurs when predators take prey at a rate proportional to the number of encounters with prey.
    Answer: Type I response

Pages: 512

11. Involved in the Type III response, a predator can turn to a/an _____.
    Answer: alternate prey

Pages: 513

12. Numerical responses of predators might include an increase in _____ or _____.
    Answer: natality, mortality, emigration (any two)

Pages: 514

13. The maximum rate of energy gain requires a/an _____.
    Answer: optimal foraging strategy

Pages: 515

14. The marginal value theorem attempts to obtain _____.
    Answer: foraging efficiency

Pages: 519

15. When an animal must decide to go back to a patch or visit a new patch, the foraging is _____ foraging.
    Answer: risk-sensitive

Matching

Choose the item from Column 2 that best matches each item in Column 1.

Pages: 508

1. Column 1: Predators become more numerous by increased reproduction in response to high grey population.
   Column 2: Numerical response

Pages: 510

2. Column 1: May stabilize prey populations
   Column 2: Type III response

Pages: 508

3. Column 1: Assumed in simple predation models
   Column 2: Type I response

Pages: 508

4. Column 1: Number of prey rises at a decreasing rate then levels off
   Column 2: Type II response

Pages: 505

5. Column 1: Handling time is a dominant component
   Column 2: Type II response

Essay

Write your answer in the space provided or on a separate sheet of paper.

Pages: 502-504

1. What are the Lotka-Volterra and Nicholson-Bailey models and why do they predict responses?
    Answer: The Lotka-Volterra and Nicholson-Baily models of predation involved paired logistic equations modified by coefficients for interactions of the two populations. Both emphasize effect of predator on prey population. Both predict oscillations in predator and prey populations. They assume random search, random encounters of predators, proportion of encounters are constant at all prey densities, and number of prey taken are in direct proportions to the number of predators. These simplified assumptions weaken the value of the models.

Pages: 505

2. Why is the Rosenweig-MacArthur model more applicable than the Lotka-Volterra and Nicholson-Bailey models?
    Answer: Although parts of the model are similar to the Lotka-Volterra and Nicholson-Bailey models, this model incorporates a prey refuge. The predator population declines from lack of food decline while prey increase in the refuge. As prey abundance increases, predators increase. This model more correctly mimics natural predator-prey interactions.

Pages: 509

3. What are Type I, II, and III responses in predation?
    Answer: In Type I, the number of prey taken per predator increase linearly to a maximum as prey density increases. In Type II, the number of prey rises at a decreasing rate toward a maximum. In Type III, the number of prey taken is low at first, then increases rapidly to an asympote, resulting in a sigmoid pattern.

Pages: 512

4. How are search image and switching related?
    Answer: When a prey becomes scarce a new species must be selected and switching to the new prey must be done. This requires a new prey image to be developed and predation will be slow on the new prey until this occurs.

Pages: 515-519

5. What significant factors and relationships influence foraging?
    Answer: Both foraging and foraging strategies are important as switching and the development of search images improve. The goal is to obtain an optimal diet by foraging techniques with regard to predator safety and risk sensitive foraging.

Multiple-Choice

Choose the one alternative that best completes the statement or answers the question.

Pages: 522
1. Herbivory includes
   A) browsing deer.
   B) acorn-eating insects.
   C) fruit-eating bats.
   D) all of the above
   Answer: D

Pages: 522
2. Grazing
   A) is never beneficial to the plants.
   B) can increase the fitness of grass.
   C) always involves fruit eating.
   D) all of the above
   Answer: B

Pages: 523
3. Some plant predators such as aphids
   A) always kill the plant.
   B) suck the sap.
   C) eat the leaves.
   D) all of the above.
   Answer: B

Pages: 523
4. For grasses moderate grazing
   A) decreases the fitness.
   B) decreases the biomass.
   C) increases the productivity.
   D) does not change the biomass.
   Answer: C

Pages: 524
5. Primary production
   A) has a pronounced effect on fitness of herbivores.
   B) has a pronounced effect on herbivore biomass.
   C) has a pronounced effect on herbivore production.
   D) all of the above
   Answer: D

Pages: 524
6. Secondary compounds in some plants
   A) do not affect animals only other plants.
   B) do not affect any animals.
   C) serve as reproductive clues for some animals.
   D) none of the above
   Answer: C

Pages: 525
7. Plant defense
    A) does not include chemical defenses.
    B) does not include structural defenses.
    C) includes mimicry.
    D) includes moving about.
    Answer: D

Pages: 525
8. Some special defenses are
    A) some hard coated seeds and fruits.
    B) spines and thorns.
    C) tough leaves and epidermal hairs.
    D) all of the above
    Answer: D

Pages: 527
9. Chemical defenses
    A) cost the plant little.
    B) could be metabolic products directly related to metabolism.
    C) are present in all plants.
    D) require a tradeoff between defense and reproduction.
    Answer: D

Pages: 527
10. Chemical defenses are
    A) proteins.
    B) sugars.
    C) amino acids.
    D) phenolics.
    Answer: D

Pages: 528
11. Inhibitors may act as
    A) warning odors.
    B) repellants.
    C) direct poisons.
    D) all of the above
    Answer: D

Pages: 528
12. Apparent plants are large, easy to locate, and
    A) usually herbaceous and short lived.
    B) they possess the cheapest type of defense-dosage independent.
    C) produce tannins and resins concentrated in bark or seeds.
    D) some are algae.
    Answer: C

Pages: 529
13. Unapparent plants
    A) are usually woody, tree like plants.
    B) produce a highly toxic cardiac glycoside.
    C) produce a dosage dependent substance.
    D) some are algae.
    Answer: B

Pages: 529
14. The main mechanism of defense against chemical substances by herbivores is
    A) excretion.
    B) secretion.
    C) storage in body fats.
    D) detoxification.
    Answer: D

Pages: 532
15. Plant growth
    A) is a function of biomass.
    B) has a consumption per biomass relationship for herbivores.
    C) can equal herbivore consumption.
    D) all of the above
    Answer: D

True-False

Write T if the statement is true and F if the statement is false.

Pages: 522
1. Predation on plants by herbivores includes only defoliation.
   Answer: False

Pages: 522
2. Continued grazing does not kill plants.
   Answer: False

Pages: 522
3. Seed predation has no impact.
   Answer: False

Pages: 522
4. Plants respond to defoliation only slowly.
   Answer: False

Pages: 523
5. Some plant predators such as aphids, tap plant juices on new growth rather than consume plant tissue.
   Answer: True

Pages: 523
6. For grasses severe grazing causes an increase in biomass.
   Answer: False

177

7. Plants supporting herbivores have a profound effect on herbivore fitness.
   Answer: True

8. Secondary compounds have a profound reproductive effect on voles.
   Answer: True

9. Plant mimicry occasionally occurs.
   Answer: True

10. Structural defenses are not significant in plant protection.
    Answer: False

11. Leaves of species of Passiflora look like inedible species and are mimics.
    Answer: True

12. Predator satiation allows a percentage of seeds or fruits to escape predation.
    Answer: True

Short Answer

Write the word or phrase that best completes each statement or answers the question.

1. Defoliation and consumption of _____ and _____ are done by herbivores.
   Answer: fruits, seeds   (any order)

2. Some plant predators acting as _____ suck the plant juices but do not kill the plant.
   Answer: parasites

3. Loss of foliage by _____ can result in the death of the plant or in a reduction of its competitive ability and fitness.
   Answer: grazers

4. Defoliation of _____ trees results in their death because they don't have the physiological traits that allow them to recover by forming new needles.
   Answer: coniferous

5. For grasses, moderate grazing typically increases _____.
   Answer: biomass production

Pages: 523

6. In studies of the scarlet gilia, grazing caused a threefold increase in flowers, fruits, and seeds, thereby giving a 2.4 fold increase in _____ over ungrazed plants.
   Answer: fitness

Pages: 525

7. Because the passion flower butterfly lays eggs on the passion flowers but will not do so if other eggs are present, the passion flower has evolved glandular outgrowths resembling insect eggs and the butterfly does not lay eggs on these plants because of _____.
   Answer: mimicry

Pages: 527

8. Seeds can be produced in such abundance that seed predators cannot eat them all
   Answer: predator satiation

Pages: 527

9. A major plant defense is _____ defense.
   Answer: chemical

Pages: 531

10. Growth of vegetation can be described by a/an _____ involving plant biomass.
    Answer: logistics equation

Matching

Choose the item from Column 2 that best matches each item in Column 1.

Pages: 525
1. Column 1: A mechanical defense of plants.
   Column 2: tough leaves and spines

Pages: 525
2. Column 1: A defense because the plant looks like another plant.
   Column 2: mimicry

Pages: 527
3. Column 1: Chemical found in fruits and seeds which is inhibitory to predators.
   Column 2: tannins

Pages: 524
4. Column 1: Chemicals in some grasses which initiate reproductive behavior in voles.
   Column 2: estrogenic hormones

Pages: 527
5. Column 1: An ubiquitous inhibitor in plant tissue.
   Column 2: phenols

Essay

Write your answer in the space provided or on a separate sheet of paper.

Pages: 522

1. What are some different ways plant predation can occur?
   Answer: Plant predation can occur as a result of fruit and seed eating, eating of foliage stems and roots, and sucking of plant juices.

Pages: 522

2. What are the responses of plants to herbivory?
   Answer: If removal of plant parts is not too extreme, the tissues regenerate and increase, but if tissue removal is too extreme, permanent damage or death will result.

Pages: 523

3. Deciduous woody plants and conifers do not respond as well as grasses to herbivory. Why?
   Answer: In conifers, if all the needles are removed the plant dies. In woody plants defoliation drains nutrients, reduces growth, and often growth forms. Grasses and some herbaceous plants regenerate easier.

Pages: 525,526

4. How do the four plant defenses work?
   Answer: Mimicry appears in plants in which leaf shapes of a sensitive to herbivory plant resemble other leaf shapes which are not sensitive. Small glandular outgrowths mimic insect eggs to discourage egg layering. Structural defenses of plants include tough epidermal layers, spines, or other structures. Chemical defenses include phenols and tannins found in seeds and fruits as well as cyanogens. Satiation occurs when large numbers of seeds occur all at once allowing some to escape predation.

CHAPTER 25 Herbivore-Carnivore Systems

Multiple-Choice

Choose the one alternative that best completes the statement or answers the question.

Pages: 536
1. Herbivory
   A) does not have the biomass of predation in the community.
   B) refers to meat eaters.
   C) supports carnivores.
   D) all of the above
   Answer: C

Pages: 536
2. Which of the following are not prey defenses.
   A) chemical
   B) cryptic coloration
   C) mimicry
   D) normal behavior
   Answer: D

Pages: 538
3. Hiding from a predator is accomplished by
   A) Batesian mimicry.
   B) cryptic coloration.
   C) warning coloration.
   D) Mullerian mimicry.
   Answer: B

Pages: 537
4. Some characteristics which are not defenses are
   A) armor.
   B) weapons.
   C) cryptic coloration.
   D) soft bodies.
   Answer: D

Pages: 538
5. Behavioral defenses do not include
   A) mimicry.
   B) distraction display.
   C) attack-abatement effect.
   D) alarm calls.
   Answer: A

Pages: 539
6. Predator satiation involves
   A) bad tasting prey.
   B) the timing of reproduction.
   C) when predators are not hungry.
   D) all of the above
   Answer: B

Pages: 540

7. For prey faced with generalized predators
   A) synchronous reproduction.
   B) satiation.
   C) asynchronous reproduction may be the best strategy.
   D) none of the above
   Answer: C

Pages: 540

8. The 13-year and 17-year cyclic appearances of the periodical cicada result in the appearance of huge numbers of progeny all at one time. Based on this evidence many survive due to
   A) mimicry.
   B) behavior adaptations.
   C) cryptic coloration.
   D) satiation.
   Answer: D

Pages: 540

9. Ambush hunting involves lying in wait for prey. It is used by
   A) frogs.
   B) lizards.
   C) alligators.
   D) all of the above
   Answer: D

Pages: 541

10. Cannibalism
    A) is not found where prey are scarce and predators frequent.
    B) is not very common.
    C) is found mainly in stressed populations.
    D) occurs in all predators.
    Answer: C

Pages: 542

11. Intraguild predation is
    A) between herbivores.
    B) between herbivores and carnivores.
    C) between two carnivorous species that feed on the same prey.
    D) when herbivore feed on carnivores.
    Answer: C

Pages: 542,543

12. The classic 10 year cycles of the snowshoe hare and the Canadian lynx in North America has been studied extensively and is now thought to be
    A) predator-prey interaction.
    B) related to sunspot activity.
    C) unrelated to tree rings.
    D) not related to annual snow accumulation.
    Answer: B

Pages: 546

13. A role for natural predation is that it acts as a regulator of prey populations and is
    A) true as studies of the lynx and hare support this statement.
    B) true if generalist predators can regulate prey populaitons by shifting their prey.
    C) is untrue if predation is independent of the ration of prey to predator.
    D) all of the above
    Answer: B

Pages: 547

14. A form of highly selective and intensive predation is that of humans. This predation
    A) is independent of prey density or predator density.
    B) results in over harvesting.
    C) along with removal of habitat has resulted in serious decline or extermination of many
       species.
    D) all of the above
    Answer: D

Pages: 548

15. The objective of regulated exploitation of a population is
    A) a sustained yield.
    B) a large biomass.
    C) extermination.
    D) a large standing crop.
    Answer: A

Pages: 549

16. The sustained yield obtained by considering many factors is
    A) a maximum sustained yield.
    B) a minimal sustained yield.
    C) an optimal sustained yield.
    D) a false sustained yield.
    Answer: C

Pages: 550

17. In populations of K-strategists the maximum rate of harvest depends on
    A) age structure.
    B) frequency of harvest.
    C) fluctuations in environment and fecundity.
    D) all of the above
    Answer: D

Pages: 552

18. An approach to management of exploitable populations is
    A) occasional yield management.
    B) temporary yield management.
    C) variable yield management.
    D) fixed quota management.
    Answer: D

Pages: 553
19. Some approaches to management of exploitable populations are
   A) harvest effort.
   B) dynamic pool model.
   C) fixed quota management
   D) all of the above
   Answer: D

Pages: 555
20. In the whaling industry
   A) whales are a drastically reduced population.
   B) whales are overexploited.
   C) if the solution of killing wolves to have more moose to hunt were applied to whales perhaps we should sink whaling boats until populations build up again.
   D) all of the above
   Answer: D

True-False

Write T if the statement is true and F if the statement is false.

Pages: 536
1. In the predator-prey relationship the prey have no defenses.
   Answer: False

Pages: 536
2. Carnivores have a serious lack of quality in their food.
   Answer: False

Pages: 536
3. The prey have only one choice and that is to run.
   Answer: False

Pages: 536
4. The predator can sit and wait or stalk and track or search and pursue.
   Answer: True

Pages: 536
5. The chemical defenses of anthropods includes cyanogenic glycosides.
   Answer: True

Pages: 536
6. Warning coloration and mimicry are not effective in protecting prey and are not commonly used.
   Answer: False

Pages: 539
7. The timing of reproduction so that many offspring are produced all at once is a defense mechanism called satiation.
   Answer: True

Pages: 540
8. Other than pursuit, predators do not have prey catching strategies.
   Answer: False

9. Cannibalism can under some conditions control the predator population.
Answer: True

10. Intraguild predation refers to predators which kill and eat species using the same resource.
Answer: True

11. In aggressive mimicry, the predator resembles the prey.
Answer: True

12. The cyclic nature of the arctic hare and the Canadian lynx in tandem is an example of predator-prey interaction.
Answer: False

13. The objective of a managed population is to obtain maximum yield.
Answer: False

14. The harvest effort method is used to establish seasons for sport hunting and fishing.
Answer: True

15. The dynamic pool method assumes animals removed will not replace natural mortality. As a result, the goal is to reduce the population so it cannot be successfully hunted.
Answer: False

Short Answer

Write the word or phrase that best completes each statement or answers the question.

1. Herbivory supports _____.
Answer: carnivory

2. The fitness of the prey depends on it's ability to elude _____ and of a herbivore to overcome _____ defenses.
Answer: predation, plant (in order)

3. _____ is a means by which snakes capture prey but it also protects snakes from enemies.
Answer: Venom

4. There are a large number of chemical defenses including the release of a/an _____ by fish.
Answer: pheromone

5. The prey can either _____(name three) when hunted.
   Answer: hide, run, fight   (any order)

6. The predator can either _____(name three) when hunting.
   Answer: sit and wait, trap, or pursue

7. Animals that have _____ and pronounced toxicity tend to have bold warning coloration such as the bright orange of the monarch butterfly.
   Answer: chemical defenses

8. _____ mimicry is the resemblance of an edible species to an inedible one, the model.
   Answer: Batesian

9. _____ refers to colorations, shapes, and behaviors which make the prey less visible.
   Answer: Cryptic coloration

10. Clams and armadillos use a defense of _____.
    Answer: armor

11. _____ predation is basically cannibalism.
    Answer: Intraspecific

12. Cycles of predator-prey populations are more extreme in _____ and poorly understood.
    Answer: boreal regions

13. The desired result of limited exploitation of a regulated government is _____.
    Answer: sustained yield

14. Exploiters experience _____ when a population is over-exploited.
    Answer: decreased catch

15. The level of sustained yield at which the population declines if it is exceeded is called the _____.

    Answer: maximum sustained yield

Matching

Choose the item from Column 2 that best matches each item in Column 1.

1. Column 1: It looks like a toxic organism but it is not.
   Column 2: Batesian mimicry

Pages: 536

2. Column 1: They are usually highly colored.
   Column 2: warning coloration

Pages: 536

3. Column 1: It looks like the environment.
   Column 2: cryptic coloration

Pages: 536

4. Column 1: The monarch butterfly feeds on milkweeds for this.
   Column 2: chemical defense

Pages: 536

5. Column 1: Secretions which ooze from millipedes.
   Column 2: chemical defense

Essay

Write your answer in the space provided or on a separate sheet of paper.

Pages: 536-539

1. Chemical defenses, behavior, and cryptic coloration are common prey defenses. How effective are these defenses in individual cases?
   Answer: Chemical defenses are widespread and include venomous snakes, pheromones in fish, secretions of arthropods, amphibians, and snakes which repel predators. Other secretions in stinkbugs and skunks are effective. Behavioral defenses run a full gamut, some as simple as warning calls are used by many species. Distraction displays are used by birds and others. Crowding together is common. Cryptic colorations involve displays, postures, etc., which tend to make the prey invisible. It is common in fish, snakes, amphibians, grasshoppers, and many other animals including infantry soldiers.

Pages: 540

2. What are the predator offenses and how effective are these?
   Answer: Predators have three ways of hunting: ambush, stalking and tracking, and search and pursuit. Using these ways predators must make decisions and have successful foraging.

Pages: 537,538

3. How do Batesian and Mullerian mimicry differ?
   Answer: In Mullerian mimicry the mimic and the model are toxic or unpalatable. In general they are usually in the same genus of taxonimic group suggesting a common evolutionary development. In Batesian mimicry, an edible species is the mimic and the inedible one the model.

Pages: 541

4. What initiates cannibalism? Is cannibalism a significant regulator of predator popoulations?
   Answer: Cannibalism is a special case of predation where the predator kills and feeds on his own species. Because of this the removal of one individual decreased the population at the same time it removes prey. Cannibalism has been found mainly in stressed populations. Cannibalism can be severe enough to decrease populations.

Pages: 554
5. What are some signs of overexploitation?
   Answer: Overexploitation is characterized by a decreasing catch per unit effort, and also by a decreasing proportion of pregnant females and a high proportion of non-reproducing individuals.

CHAPTER 26 Parasitism

Multiple-Choice

Choose the one alternative that best completes the statement or answers the question.

Pages: 558
1. Parasites
   A) do not harm the host.
   B) do not live in the host just on the outside.
   C) do not cause diseases.
   D) obtain nourishment from the host.
   Answer: D

Pages: 558
2. Parasites do not include
   A) viruses.
   B) bacteria.
   C) green plants.
   D) some invertebrates.
   Answer: C

Pages: 558
3. Recent studies indicate parasites
   A) increase death rates.
   B) decrease death rates.
   C) increase susceptibility to predation.
   D) all of the above
   Answer: C

Pages: 558
4. Macroparasites
   A) are microscopic.
   B) are all endoparasites.
   C) are relatively large (parasitic worms, mites, etc.).
   D) all of the above
   Answer: C

Pages: 558
5. Microparasites
   A) are large.
   B) have a short generation time.
   C) have very little effect on the host.
   D) are animals like fleas, ticks, and worms.
   Answer: B

Pages: 558
6. Ectoparasites
   A) are found inside the host.
   B) do not affect the host.
   C) are inside plants.
   D) are on the outside in the hair and feather of hosts.
   Answer: D

Pages: 562

7. The brainworm of the white tailed deer
   A) is ingested as a larvae while infecting a snail.
   B) moves to the brain where as an adult it lays eggs.
   C) eggs hatch in the lungs and larvae exit the host.
   D) all of the above
   Answer: D

Pages: 563

8. Indirect transmission
   A) refers to parasites that require different hosts for different stages.
   B) occurs with the brainworm in deer.
   C) occurs with the black grub.
   D) all of the above
   Answer: D

Pages: 560

9. Direct transmission is
   A) the carrying of the parasite from one host to another by direct contact.
   B) the carrying of the parasite from one host to another by a vector.
   C) the carrying of the parasite from one host to another by carrier.
   D) all of the above
   Answer: D

Pages: 561

10. Which of the following is an example of direct transfer:
    A) Lice are transmitted from one hog to another by rubbing.
    B) black spot minnows by larvae.
    C) tapeworm larvae in carnivores.
    D) all of the above
    Answer: A

Pages: 565

11. The rapid spread of viral and bacterial diseases in dense populations of animals are called
    A) epizootics.
    B) epidemics.
    C) invasion.
    D) all of the above
    Answer: A

Pages: 566

12. After infection, the host, depending upon the parasite
    A) may have an immune response.
    B) may have abnormal growths.
    C) may form encrustation.
    D) all of the above
    Answer: D

Pages: 561
13. A form of plant parasitism by photosynthetic plants is
    A) hemiparasitism.
    B) holoparasitism.
    C) larval parasitism.
    D) all of the above
    Answer: A

Pages: 589
14. Parasitism as a mechanism of prey regulation depends on transmission by
    A) direct means in highly scattered populations.
    B) indirect means with poorly adapted alternate hosts.
    C) direct means in herds or close contact.
    D) none of the above
    Answer: C

Pages: 560
15. Which of the following is true:
    A) As host immunity increases, parasite numbers increase.
    B) In low host populations parasite spreading by indirect transmission is common.
    C) Rabies produces long term immunity.
    D) All parasites produce diseases.
    Answer: B

Pages: 571
16. Parasite population sizes are not
    A) related to density of host population.
    B) related to distribution of hosts body.
    C) not related to parasite competition.
    D) none of the above
    Answer: D

Pages: 573
17. Social parasitism refers to
    A) brood parasitism.
    B) endoparasitism.
    C) a rabbit viral parasite.
    D) parasites of long duration.
    Answer: A

True-False

Write T if the statement is true and F if the statement is false.

Pages: 558
1. Parasitism is a condition in which two organisms live together both deriving nourishment.
   Answer: False

Pages: 558
2. Parasitoids appear as if they are parasites.
   Answer: False

Pages: 558
3. Parasites can increase death rates, decrease birth rates, and decrease growth rates.
Answer: True

Pages: 558
4. Ectoparasites are internal parasites.
Answer: False

Pages: 558
5. Macroparasites are small internal parasites such as bacteria.
Answer: False

Pages: 559
6. Hosts are food and habitat, all habitats are exploited by some parasite including blood, brain, and skin of animals.
Answer: True

Pages: 559
7. A major problem for parasites is gaining access to the host and an exit from the host.
Answer: True

Pages: 560
8. The definitive host is one in which the parasite becomes an adult.
Answer: True

Pages: 560
9. Direct transfer involves several life forms to transmit the parasite from one host to another.
Answer: False

Pages: 561
10. Ascaris or roundworm, a nematode, is a typical microparasite.
Answer: False

Pages: 562
11. The brainworm is an indirectly transferred endoparasite of the white-tailed deer.
Answer: True

Pages: 567,568
12. Host responses to parasite infections can be biochemical, abnormal growths, sterility, behavioral changes, or changes in mate selection.
Answer: True

Pages: 569
13. The effect of parasites on host populations is negligible.
Answer: False

Pages: 569
14. For macroparasites to persist, they must find transmission patterns dependent on food webs of more than one species because they are usually indirectly transferred.
Answer: True

15. Social parasitism such as kleptoparasitism are common in behaviors such as the eagle when it steals prey from the osprey.
Answer: True

Short Answer

Write the word or phrase that best completes each statement or answers the question.

Pages: 558

1. _____ are organisms which live with a host from which they derive nourishment but don't kill the host.
Answer: Parasites

Pages: 558

2. _____ are insects which live on a host and consume the host during the larval stage thus killing the host.
Answer: Parasitoids

Pages: 558

3. Parasites include _____(name four).
Answer: viruses, bacteria, fungi, invertebrates (any order)

Pages: 558

4. A/an _____ occurs when a plant or animal deviates from normal well-being.
Answer: disease

Pages: 558

5. Parasites can be _____and live outside the host's body.
Answer: ectoparasites

Pages: 558

6. A/an _____ is a parasite which lives inside the host's body.
Answer: endoparasite

Pages: 558

7. Bacteria, viruses, and protozoans can be parasites classed by size called _____.
Answer: microparasites

Pages: 558

8. Parasites such as worms, flukes, fleas, etc. are classed by size called _____.
Answer: macroparasites

Pages: 559

9. Hosts are the _____ of parasites and many have been exploited by various parasites.
Answer: habitat

Pages: 559

10. The chief problems of parasites is gaining _____ and _____ of the host.
Answer: access, escape (any order)

Pages: 561
11. The debilitating parasites of animals and man (lice, ticks, fleas) are spread by _____.
    Answer: direct contact

Pages: 561
12. Mistletoes are _____ because they draw water and minerals from the host and they reduce the growth of the host.
    Answer: hemiparasites

Pages: 562
13. Parasites spread by intermediate hosts have adapted a stage in the _____ to accomplish the transfer.
    Answer: life cycle

Pages: 565
14. A plant fungus with multiple stages on two plants (tree and gooseberry) is the _____.
    Answer: white pine blister rust

Pages: 568
15. A host may respond to the parasite by _____ (name three) and in birds mate selection.
    Answer: biochemical responses, abnormal growths, behavior changes   (any order)

Pages: 565
16. _____ are to animal populations as epidemics are to humans.
    Answer: Epizootics

Pages: 573
17. Social types of parasitism are: _____ (name three).
    Answer: social parasitism (dependent on social organization), brood parasitism (incubation of eggs), kleptoparasitism (stealing eggs or young)   (any order)

Matching

Choose the item from Column 2 that best matches each item in Column 1.

Match the parasite with its proper classification.

Pages: 561
1. Column 1: Nematode
   Column 2: endoparasite

Pages: 559
2. Column 1: Tick
   Column 2: ectoparasite

Pages: 573
3. Column 1: Rabbit viral parasite
   Column 2: endoparasite

Pages: 559
4. Column 1: Flea
   Column 2: ectoparasite

Pages: 558

5. Column 1: Tapeworm
   Column 2: endoparasite

Essay

Write your answer in the space provided or on a separate sheet of paper.

Pages: 566

1. What is parasitism? How many kinds are present in terms of funtioning parasites?
   Answer: Parasitism is defined as a condition in which two animals live together with one deriving food from the body of the other. Parasites can be classed according to where the parasite lives. Ectoparasites are found on external surfaces and endoparasites are found in the body where different parasites attack different parts of the body. They can also be classed on size; macroparasites are large like fleas and mites; microparasistes are small including viruses and bacteria.

Pages: 573

2. What is the difference between parasitism and social parasitism?
   Answer: Parasitism refers to a condition where the parasite obtains food directly from the host body. Social parasitism refers to conditions where food is stolen or taken from the host or parasites young are placed in host's care.

Pages: 561,562

3. How are parasites transmitted to new hosts?
   Answer: There are two ways parasites are transmitted: direct transmission from host to host with physical contact and indirect transmission from an alternate host carrying some other stage in the life cycles. In general, microparasites are directly transmitted and macroparasites can be directly or indirectly transmitted.

Pages: 569

4. How do parasites influence host populations?
   Answer: Parasite regulation is dependent on the nature of transmission, density, and dispersion of the host. In clumped or crowded areas, direct transmission by parasites can exert a considerable reduction in host organisms. Viral distemper in racoon has a regulatory effect. Mountain sheep populations are controlled by pneumonia as a result of high lung worm infection.

Pages: 560

5. A definitive host differs from an intermediate host. How is this difference important in parasitism?
   Answer: The definitive host is the host in or on which the parasite reaches the adult stage. An intermediate host may harbor some development phase or all other phases but not the adult. A parasite may have several intermediate hosts but only one adult host. Some parasites are highly specific in having several intermediate hosts.

Multiple-Choice

Choose the one alternative that best completes the statement or answers the question.

Pages: 582
1. Mutualism can occur
   A) between two populations.
   B) after coevolution.
   C) in absence of coevolution.
   D) all of the above
   Answer: D

Pages: 582
2. Mutualism is not
   A) symbiotic.
   B) nonsymbiotic.
   C) facultative.
   D) without benefits.
   Answer: D

Pages: 584
3. Obligate nonsymbiotic mutualisms include
   A) ectomycorrizae.
   B) indomychorrizae.
   C) fungus growing ants.
   D) all of the above
   Answer: C

Pages: 587
4. Defensive mutualism includes
   A) a fungus on grass which makes the grass unpalatable for cattle.
   B) endomycorrhizae.
   C) indirect mutualisms.
   D) all of the above
   Answer: A

Pages: 587
5. Pollination by mutualists
   A) occurs when plants are rare.
   B) occurs when plants are scarce.
   C) does not occur with wind pollination.
   D) all of the above
   Answer: D

Pages: 587
6. Species of yucca are pollinated by species of yucca moths.
   A) This is an example of separate evolution.
   B) This does not cost the yucca plants energy.
   C) The relationship is a nonsymbiotic obligate mutualism.
   D) Both species can occur without the other.
   Answer: C

7. In the golden bee-orchid mutualism
   A) results in pollination of the orchids.
   B) orchids are far apart and bees must travel long distances.
   C) some have flowers which mimic female bees and when a male bee attempts to copulate pheromones is attached.
   D) all of the above
   Answer: D

8. There are many nectar-feeding animals
   A) which exploit plants.
   B) most do not spread pollen.
   C) so plants are not required to expend energy in attracting insects for nectar and fragrances.
   D) but the relationship is not mutualism.
   Answer: A

9. Many species such as blackberries, elderberries, cherries, and goldenrods
   A) do not attract individual species of obligate pollinators.
   B) do not attract diversity pollinators due to over production.
   C) flower profusely and provide a glut of nectar.
   D) are specialists.
   Answer: C

10. Seed dispersal is commonly a mutualistic act such as
    A) the Clark nutcracker which disperses whitebark pine where the nutcracker is a seed eater.
    B) chipmunks which also store and eat pinion pine.
    C) myrmecochores which use ants to store and disperse seeds.
    D) all of the above
    Answer: D

11. In nature, seed dispersal requires that evolution produce
    A) fruits exploited by large numbers of animals.
    B) fruits spread by specific animals which are exclusive consumers.
    C) both A and B
    D) Neither A or B, seed dispersal is best served by a small number of birds and mammals for each fruit.
    Answer: C

12. The origin of mutualism is
    A) well known.
    B) probably by diffuse evolution.
    C) by a gene for gene selection of dual characteristics.
    D) is random.
    Answer: B

Pages: 595

13. In order to demonstrate the advantages of mutualism it is necessary to show
    A) increased survival of two populations together vs the two populations apart.
    B) an increased survival of either population together vs apart.
    C) no effect together vs apart for survivorship.
    D) all of the above
    Answer: A

Pages: 595

14. In studies of the mathematics of mutualism
    A) Lotka-Volterra equiations for competition are used.
    B) the signs of the Lotka-Volterra equation for alpha are made positive.
    C) to cap growth and stabilize equations a third species that is a predator or competitor of one
       of the mutualists must be brought in to the equation.
    D) all of the above
    Answer: D

True-False

Write T if the statement is true and F if the statement is false.

Pages: 582

1. Mutualism is when two organisms live together and obtain their food from each other.
   Answer: False

Pages: 582

2. Mutualism came about exclusively by coevolution.
   Answer: False

Pages: 582

3. Another model for the origin of mutualism states that organisms evolved in different types of
   habitats and when invaded by new habitats they adjusted.
   Answer: True

Pages: 582

4. Mutualism may be symbiotic, living together or nonsymbiotic, living apart. What mutualists do
   is the issue.
   Answer: True

Pages: 582

5. Mutualism can be facilitative: not necessary for survival of either organism or obligate:
   necessary for the survival of one of the organisms.
   Answer: True

Pages: 586

6. Endomycorrhizae are not obligate symbiotic mutualists.
   Answer: False

Pages: 587

7. Defense mechanisms are usually obligate symbiotic mutualists.
   Answer: False

8. Pollination in orchids does not do anything for the pollinator birds.
   Answer: False

9. In seed dispersal, many seeds are eaten by the disperser which do not benefit the plant; however many are dispersed to suitable sites.
   Answer: True

10. Understanding the role of mutualism and population dynamaics depends on stronger mathematical models and empirical data.
    Answer: True

Short Answer

Write the word or phrase that best completes each statement or answers the question.

1. _____ is when reciprocal selection pressures benefit two organisms and is a result of _____.

   Answer: Mutualism, coevolution   (in order)

2. _____ adaptation and counter adaptation bring about evolutionary change.
   Answer: Coevolution

3. _____ occur over many species as is the case in some plant pollinators.
   Answer: Diffuse interactions

4. Whether _____ have evolved gene for gene over time is unknown.
   Answer: mutualistic species

5. Mutualism may be _____ in which both organisms live together with at least one completely dependant on the other.
   Answer: symbiotic

6. Obligate symbiotic mutualisms include _____ and _____.
   Answer: ectomycorrizae, endomycorrhizae   (any order)

7. _____ include pollination and fungus growing ants. (be specific)
   Answer: Obligate nonsymbiotic mutualisms

8. _____ include defense mechanisms and indirect mutualism as occurs in fungal infections in grass that prevent cattle from eating. (be specific)
   Answer: Facultative mutualisms

Pages: 588-593
9. _____ and _____ are mutualisms in which a price in energy is paid by the plant and pollinator.
Answer: pollination, seed dispersal (any order)

Pages: 595
10. _____ for mutualism can be studied by the use of modified Lotka-Volterra equations for competition by making the negative alphas positive for mutualism.
Answer: Population effects

Matching

Choose the item from Column 2 that best matches each item in Column 1.

Pages: 583
1. Column 1: Endomycorrhizae
   Column 2: symbiotic obligate mutualist

Pages: 587
2. Column 1: Yucca and yucca moth
   Column 2: nonsymbiotic obligate mutualist

Pages: 583
3. Column 1: Ectomycorrhizae
   Column 2: symbiotic obligate mutualist

Pages: 588
4. Column 1: Orchids and golden bees
   Column 2: nonsymbiotic obligate mutualist

Pages: 590
5. Column 1: Seed dispersal of whitebark pine by nutcracker
   Column 2: nonsymbiotic obligate mutualist

Essay

Write your answer in the space provided or on a separate sheet of paper.

Pages: 582
1. What is mutualism? Discuss the difference between obligatory symbiotic mutualism and obligatory nonsymbiotic mutualism.
   Answer: Mutualism is when two organisms have a relationship of mutual benefit. In some cases they cannot survive without the other; this is obligate mutualism. When they live together, it is a symbiotic mutualism. A good example of symbiotic mutualism is the mychorrizal relationship between fungi which grow in or on plant roots. Mychorrizae assist in the uptake of water and minerals and the green plant supplies food. The lichen is another example. Nonsymbiotic obligatory mutualism organisms interact but don't live together. Pollination mechanisms and food dispersal mechanisms are the best examples.

Pages: 582

2. What is coevolution and how does it apply to mutualism?
    Answer: Coevolution occurs when traits in two species evolve in relation to each other or path of evolution is together. This has not been shown on a gene-for-gene relationship in mutualism.

Pages: 586

3. What are mycorrhizal mutualisms and why are they important?
    Answer: The mycorrhizae fungi growing in or on roots of plants provide increased uptake of minerals and water in exchange for food. This is a tenuous relationship dependent on environment. These relationships are important because they allow the green plant to exploit more environments that would be possible without mycorrhizae.

Pages : 590

4. Why is seed dispersal by animals important?
    Answer: By eating fruit or burying seeds, animals scatter seeds away from parent plants, dropping some in suitable habitats. Passage of seeds through digestive tract improves germination of many types of seeds.

Pages : 587

5. What is the importance of pollination to plants and animals?
    Answer: Pollination is necessary for the completion of the plant life cycle. Some plants are wind-pollinated; others depend upon animals such as bees, birds, and bats to carry pollen from plant to plant. In some cases this relationship is obligatory. Animals are also used to complete the life cycle. The flower in many cases has evolved nectar or fragrances to attract the animals and hence to pollinate the flowers.

CHAPTER 28 Community Structure

Multiple-Choice

Choose the one alternative that best completes the statement or answers the question.

Pages: 598
1. A large number of plants and animals interacting in an area is
   A) an edge.
   B) an ecotone.
   C) a community.
   D) a species diversity.
   Answer: C

Pages: 598
2. Differences in terrestrial communties are defined by
   A) number and kind of animals.
   B) form and structure of vegetation.
   C) physical environment.
   D) soil.
   Answer: B

Pages: 598
3. Raunkiaer's life form which describes perennial shoots and buds on the surface of the ground to 25 cm high are
   A) therophytes.
   B) epiphytes.
   C) geophytes.
   D) chamaephytes.
   Answer: D

Pages: 598
4. A community with a high percentage of perrennial tissue well above the ground will be characteristic of
   A) warm climates.
   B) arctic climates.
   C) northern climates.
   D) all of the above
   Answer: A

Pages: 599
5. Vertical stratification
   A) is determined by the form of the plants.
   B) provides the structure for animal life.
   C) is best developed in forests.
   D) all of the above
   Answer: D

Pages: 599
6. The _____ is the site of primary productivity.
   A) therophytes
   B) canopy
   C) shrubs
   D) understory
   Answer: B

Pages: 601
7. Horizontal structure is present as
   A) large patches spatially separated from one another.
   B) heterogeneity between patches.
   C) heterogeneity within patches.
   D) all of the above
   Answer: D

Pages: 602
8. A dominant species in a forest community is
   A) a canopy species.
   B) a shrub.
   C) a herb.
   D) all of the above
   Answer: A

Pages: 602
9. A dominant organism
   A) has a low importance index.
   B) has little direct influence.
   C) is one whose presence is critical to the community.
   D) is rare.
   Answer: C

Pages: 602
10. Species diversity is determined
    A) from species richness (number of species).
    B) from evenness (relative abundance).
    C) by the Shannon index.
    D) all of the above
    Answer: D

Pages: 602
11. Species diversity
    A) increases from the tropics to the arctic.
    B) increased from the ocean to ocean depths.
    C) increases from east to west in North America.
    D) stays the same throughout the world.
    Answer: C

Pages: 604

12. An historical hypothesis that explains diversity is
    A) climatic stability hypothesis.
    B) productivity hypothesis.
    C) dynamic equilibrium hypothesis.
    D) ecological time hypothesis.
    Answer: D

Pages: 607

13. The random niche model
    A) explains diversity.
    B) explains the strata hypothesis.
    C) explains species abundance.
    D) edges communities.
    Answer: C

Pages: 608

14. An ecotone is
    A) an ecological theory.
    B) a clear zone.
    C) a patch.
    D) a transitional zone between two communities.
    Answer: D

Pages: 611

15. A rule of thumb for island biogeography is
    A) islands have little life.
    B) islands are too small for large numbers of species.
    C) a tenfold increase in area results in a doubling in species.
    D) islands have only exotic species.
    Answer: C

Pages: 614

16. Essentially island habitat
    A) can be a bog.
    B) can be a pond.
    C) can be a dune.
    D) all of the above
    Answer: D

Pages: 616

17. Clearing of forested land in eastern U.S.
    A) left patches of woodlots.
    B) increased species diversity.
    C) produced parks and recreational areas.
    D) all of the above
    Answer: D

Pages: 620
18. A cornerstone of community structure is
   A) mutualism.
   B) neutralism.
   C) commensualism.
   D) competition.
   Answer: D

Pages: 621
19. The effect of _____ is easily demonstrated.
   A) predation
   B) neutralism
   C) parasitism
   D) all of the above
   Answer: A

Pages: 625
20. Community patterns are conceived as
   A) continuums along gradients.
   B) discrete entities.
   C) chance events.
   D) random events.
   Answer: A

Pages: 629
21. Methods of analyzing communities are
   A) community ordination and principle components analysis.
   B) discrete areas analysis.
   C) unit analysis.
   D) an unorganized analysis or random analysis.
   Answer: A

Pages: 625
22. Concepts of communtities are
   A) organismic or individualistic.
   B) competitive or noncompetitive.
   C) random or chance.
   D) all of the above
   Answer: A

Pages: 608
23. An induced edge is
   A) a result of disturbances.
   B) developmental.
   C) is permanent.
   D) is temporary.
   Answer: A

205

Pages: 604

24. _____ influences plant species diversity.
    A) Type of soil
    B) Nutrients
    C) Climate
    D) all of the above
    Answer: D

Pages: 600

25. The primary producers in a lake are found in the
    A) epilimnion.
    B) hypolmnion.
    C) thermocline.
    D) all of the above
    Answer: A

True-False

Write T if the statement is true and F if the statement is false.

Pages: 598
1. Therophytes are perennials.
   Answer: False

Pages: 598
2. Chamaephytes grow on the desert.
   Answer: True

Pages: 598
3. Differences in terrestrial communities are defined by animal types and phyla.
   Answer: False

Pages: 599
4. A distinctive feature of many communities is vertical stratification.
   Answer: True

Pages: 600
5. Stratification is absent from aquatic communities.
   Answer: False

Pages: 601
6. Horizontal structure of a community is characterized by patches.
   Answer: True

Pages: 602
7. The dominant species is always the most abundant.
   Answer: False

Pages: 601
8. In the forest, patterns of light and shade shape the development of the understory.
   Answer: True

206

Pages: 602

9. Index species are indicators of dominant species.
Answer: True

Pages: 602

10. Species diversity does not consider evenness (relative abundance).
Answer: False

Pages: 605

11. Hypotheses related to climate are adequate to explain species diversity.
Answer: False

Pages: 607

12. The random niche model is one hypothesis to explain species abundance.
Answer: True

Pages: 608

13. Inherent edges are inherently unstable.
Answer: False

Pages: 609

14. Ecotones are edges with organisms.
Answer: False

Pages: 612

15. Turnover rates are greater for near islands than for distant islands.
Answer: True

Pages: 614

16. Habitat fragmentation produces an island effect.
Answer: True

Pages: 616

17. There appears to be a negative correlation between edge species of plants and size of forest islands.
Answer: True

Pages: 620

18. Population interactions which are not of importance in communities are competition and predation.
Answer: False

Pages: 621

19. Keystone species are top predators.
Answer: False

Pages: 625

20. Two opposing philosophies of communities are the organismic and the individualistic philosophies.
Answer: True

Short Answer

Write the word or phrase that best completes each statement or answers the question.

Pages: 598
1. A _____ is desribed as a naturally occurring assemblage of plants and animals living in the same environment and interacting in such a way that each affects the well-being of others.
   Answer: community

Pages: 598
2. _____ are groups of species that feed or forage in a similar way in the community.
   Answer: Guilds

Pages: 598
3. Terrestrial communtities are described by the form and structure of the vegetation called

   _____.
   Answer: life forms

Pages: 598
4. Raunkiaer's life forms are: _____(name six).
   Answer: therophytes (annuals), geophytes (buds or rhizomes), hemicryptophytes (perennial
              shoots close to the ground), phanerophytes (perennial buds in the air), epiphytes
              (growing on other plants), chamaephytes (perennial shoots on the ground)   (any
              order)

Pages: 599
5. _____ is a distinctive feature of communities.
   Answer: Vertical stratification

Pages: 601
6. The more strata a community has, the more _____ the plant and animal life.
   Answer: diverse

Pages: 602
7. As horizontal areas are explored, the plant and animal life are found in _____.
   Answer: patches

Pages: 602
8. _____ control the structure of communities.
   Answer: Dominants

Pages: 602
9. _____ are species critical to the integrity of the community.
   Answer: Keystone species

Pages: 602
10. To determine dominance ecologists measure _____(name five).
    Answer: relative abundance, relative dominance, relative frequency, relative importance, index
               species   (any order)

Pages: 602
11. _____ measures species abundance and richness.
    Answer: Species diversity

12. _____ is influenced by regional diversity, which is influenced by climatic history, accidents, and geographical position of dispersal barriers.
    Answer: Local diversity

13. The _____ proposes that diversity relates to the age of the community.
    Answer: evolutionary time hypothesis

14. The _____ proposes that time is needed for species to disperse into unoccupied areas of a suitable habitat.
    Answer: ecological time hypothesis

15. The _____ proposes that the more complex and heterogeneous the physical environment, the more diverse will be its flora and fauna.
    Answer: spatial heterogeneity hypothesis

16. Species abundance has been explained by the _____(name four).
    Answer: random niche model, niche preemption hypothesis, log-normal hypothesis   (any order)

17. The area where two different communities intergrade between each other are _____.
    Answer: ecotones

18. _____ produces variation in communities.
    Answer: Habitat fragmentation

19. _____ link one fragment to another.
    Answer: Corridors

20. Normal interactions, _____, _____, parasitism, and _____ produce the structure of communities.
    Answer: predation, competition, mutualism   (any order)

21. The two opposing concepts of community are _____ and _____.
    Answer: organismic, individualistic   (any order)

Matching

Choose the item from Column 2 that best matches each item in Column 1.

1. Column 1: Buds buried in ground on a bulb or rhizome
   Column 2: geophytes

Pages: 598
2. Column 1: Annuals
   Column 2: therophytes

Pages: 598
3. Column 1: Buds, perennials, trees, shrubs, over 25 cm tall
   Column 2: phanerophytes

Pages: 598
4. Column 1: Perennial shoots or buds up to 25 cm high
   Column 2: chamaephytes

Pages: 598
5. Column 1: Plants growing on other plants
   Column 2: epiphytes

Pages: 598
6. Column 1: Perennial shoots or buds close to the ground often covered by litter
   Column 2: hemicrytophytes

Essay

Write your answer in the space provided or on a separate sheet of paper.

Pages: 598
1. What is a community and what are its characteristics?
   Answer: A community is a naturally occurring group of plants and animals in a particular environment. The community is characterized by species composition or diversity, stratification, and life forms. It is usually named by the dominant plant species. Some feel that it is not set off from other communities but is part of a continuum along climatic and soil gradients.

Pages: 599,600
2. What is the role of vertical stratification and horizontal structure?
   Answer: Vertical stratification refers to the layering of physical habitat, whereas horizontal structure refers to patches of organisms scattered horizontally. Vertical stratification in a forest would be the canopy (dominants), sub-dominants, understory, herbaceous, and soil layers. An example of horizontal structure is the change of vegetation patches across old fields.

Pages: 602
3. What is a dominant or keystone species and how is this determined?
   Answer: A dominant species or keystone species is one that influences the community the most. Usually a species index is formulated by rating each species according to importance values obtained from relative abundance, dominance, and frequency.

Pages: 602
4. What is species diversity and why is it important?
   Answer: Species diversity is calculated from species richness and abundance. A number on indexes have been developed but the most commonly used is the Shannon index.

Pages: 612,614

5. How are fragment size, species composition and island geography related?

Answer: Large areas have a larger species composition. Therefore larger islands and larger fragments should have higher species composition values. However for islands, distance from the mainland decreased species composition. Fragmentations are parts of a larger area and islands are not, so comparisions may not be valid.

Multiple-Choice

Choose the one alternative that best completes the statement or answers the question.

Pages: 634
1. A disturbance is
   A) excess animals.
   B) a physical force resulting in mortality.
   C) human interactions.
   D) all of the above
   Answer: D

Pages: 635
2. Most fires are not
   A) underground in the mineral layer.
   B) in the canopy.
   C) in the under brush.
   D) in the litter.
   Answer: A

Pages: 642
3. An ecosystem which depends on frequent fires to maintain itself is the
   A) deciduous forest.
   B) grassland.
   C) desert.
   D) tundra.
   Answer: B

Pages: 638
4. Too frequent fires deplete the seed resource and alters the communities in the
   A) deciduous forest.
   B) tundra.
   C) chaparral.
   D) all of the above
   Answer: C

Pages: 641
5. Crown fires are most common in
   A) deciduous forests.
   B) deserts.
   C) tundra.
   D) coniferous forests.
   Answer: D

Pages: 642
6. The most damaging fire is the
   A) crown fire.
   B) litter fire.
   C) ground fire.
   D) underbrush fire.
   Answer: C

Pages: 641

7. Low intensity but frequent litter fires
    A) consume dead grass and mulch.
    B) do not kill rhizomes or roots.
    C) remove trees.
    D) all of the above
    Answer: D

Pages: 643

8. Some trees like the jack pine are fire dependent because
    A) they grow rapidly.
    B) have thick fire resistant bark.
    C) fire opens cones releasing seeds to fire prepared seed bed.
    D) all of the above
    Answer: D

Pages: 646

9. Selection cutting selects
    A) all marketable timber and removes it.
    B) 10%-70% of the stand and removes it.
    C) mature single or groups of trees and removes them.
    D) trees and cuts everything down.
    Answer: C

Pages: 647

10. The most powerful agent of disturbance in communities is
    A) animals.
    B) fire.
    C) wind.
    D) humans.
    Answer: D

Pages: 649

11. The great regenerator of nutrients is
    A) wind.
    B) fire.
    C) water.
    D) hurricanes.
    Answer: B

Pages: 651

12. A few species require the periodic fires to maintain habitat such as
    A) the Kirtland's warbler
    B) robins.
    C) field mice.
    D) all of the above
    Answer: A

Pages: 652
13. Resistance and maintenance of community stability is dependent on
    A) a large biotic structure.
    B) nutrients and energy stored in individuals.
    C) returning to original state.
    D) all of the above
    Answer: D

Pages: 634
14. Disturbances
    A) are the means which landscape diversity is obtained.
    B) is an outcome of natural events.
    C) brought about by humans.
    D) all of the above
    Answer: D

True-False

Write T if the statement is true and F if the statement is false.

Pages: 639
1. After a major disturbance the land never returns to the original communities.
   Answer: False

Pages: 641
2. Careless visitors start most of the forest fires in the Western U.S.
   Answer: False

Pages: 641
3. In closed stands of coniferous forests crown fires are common.
   Answer: True

Pages: 643
4. In the western states, lodge pole and jack pine are dependent on fire for seed germination.
   Answer: True

Pages: 641
5. In the North American prairie grass fires started by lightening occur every 20 or 30 years.
   Answer: False

Pages: 646
6. The most destructive logging practice is the selective cut.
   Answer: False

Pages: 649
7. Fire is the great regenerator of nutrients.
   Answer: True

Pages: 645
8. When deer exceed the carrying capacity some starve and the forest community may be reduced to a grassland area under the canopy.
   Answer: True

Pages: 646

9. Clear cutting maintains the forest composition.
   Answer: False

Pages: 652

10. Communities of large biomass and structure are more resistant to disturbances.
    Answer: True

Short Answer

Write the word or phrase that best completes each statement or answers the question.

Pages: 634

1. _____ is the means by which nature changes and diversity is maintained.
   Answer: Disturbance

Pages: 634

2. A disturbance is any _____, such as fire, wind, flood, extreme cold temperatures, or an epidemic which results in mortality or loss of biomass.
   Answer: physical force

Pages: 634

3. _____ is measured by the proportion of the total biomass or population removed.
   Answer: Intensity

Pages: 635

4. Intensity is related to the _____ and _____ of an event such as fire which can be small or large.
   Answer: frequency, scale   (any order)

Pages: 641

5. _____ are major disturbances feeding on the litter layer in forests and the dead grass and mulch in grasslands.
   Answer: Surface fires

Pages: 641

6. _____ burn the canopy in forest lands and are more characteristic of conifers.
   Answer: Crown fires

Pages: 642

7. _____(be specific) consume organic matter.
   Answer: Ground fires

Pages: 646

8. An absence of wind and water, a/an _____ can create
   Answer: drought

Pages: 646,647

9. _____(name three) are man-made disturbances which fragment natural communitites.
   Answer: Timber harvesting, surface mining, cultivation   (any order)

10. _____ is the tendency of a community to maintain a steady state and _____ is the ability of the system to maintain equilibrium; both of which are responses to disturbances.
Answer: Resistance, resilience   (in order)

Matching

Choose the item from Column 2 that best matches each item in Column 1.

Pages: 641
1. Column 1: Typical of grass fires
   Column 2: surface fire

Pages: 641
2. Column 1: Occurs mostly in conifers
   Column 2: crown fire

Pages: 643
3. Column 1: Does not harm thick barked trees
   Column 2: surface fire

Pages: 642
4. Column 1: Burns humus
   Column 2: ground fire

Pages: 641
5. Column 1: Kills herbs and seedlings in the forest
   Column 2: surface fire

Essay

Write your answer in the space provided or on a separate sheet of paper.

Pages: 638
1. What are gaps and why are they important?
   Answer: Small disturbances create gaps where light increases and soil is exposed. These become localized areas of regeneration and growth, open space for colonization by plants.

Pages: 644,645
2. Several forces act as agents of disturbance. Show how wind and drought might alter the environment?
   Answer: Wind can effect minor gap formation or major disturbances. It shapes the canopy of some trees, and uproots others so that succession starts again. Hurricanes have a devastating effect over large areas and regeneration occurs after they have passed. Prolonged drought can kill shallow rooted plants. Shorter drought periods can dry up wetland driving out animal species.

Pages: 641-643
3. Why is fire an important component of natural systems?
    Answer: Fire is a naturally occurring disturbance because lightning starts many fires. It regenerates fire-dependent ecosystems, thins forest stands, reduces litter buildup, and recycles nutrients.

Pages: 649
4. What are some effects of disturbances on nutrient recycling?
    Answer: Most disturbances release nutrients to recycling; cause loss of nutrients, accelerate leaching, loss from system by surface runoff, and regenerate nutrients.

Pages: 634
5. What are the characteristics of a disturbance?
    Answer: There are two characteristics of disturbances. One is intensity which is related to the extent of change in the community (a longer duration also increases change). Intensity is measured by the proportion of biomass or population removed. Frequency, the time between disturbances, is another characteristic.

CHAPTER 30 Succession

Multiple-Choice

Choose the one alternative that best completes the statement or answers the question.

Pages: 656
1. Succession
    A) occurs following a disturbance.
    B) is a static situation.
    C) is a fluctuation.
    D) does not occur in some regions.
    Answer: A

Pages: 656
2. Primary succession could start
    A) in a metropolitan park.
    B) on bare rock of a strip mine.
    C) in an abandoned pasture.
    D) all of the above
    Answer: B

Pages: 656
3. A secondary sere might include
    A) grass and herbs to brush to intermediate trees to beech-maple forest in a metro park.
    B) grass and herbs to shrubs in an abandoned pasture.
    C) shrubs in a cutover forest to forest.
    D) all of the above
    Answer: D

Pages: 656
4. When changes in the environment are brought about by organisms themselves, succession is
    A) allogenic.
    B) unidirectional.
    C) terminal and lasts centuries.
    D) autogenic.
    Answer: D

Pages: 656
5. Terrestial primary succession
    A) can start on dry sand.
    B) can start on bare rock.
    C) is initiated by pioneer species.
    D) all of the above
    Answer: D

Pages: 658
6. Primary succession on sand dunes was a classic study by
    A) Walker and associates.
    B) Wood and Del Moral.
    C) Cowles.
    D) Keevers.
    Answer: C

Pages: 659

7. Secondary successions
   A) were originally studied in old fields of the Piedmont region.
   B) can be studied along railroad tracks.
   C) are usually studied in disturbed areas.
   D) all of the above
   Answer: D

Pages: 659

8. In several terrestrial stages
   A) broomsedge is an important plant.
   B) <u>K</u>-selected species are important.
   C) there are no climaxes.
   D) the plants change but the animals do not.
   Answer: A

Pages: 660

9. Succession in aquatic environments
   A) is secondary.
   B) starts with open water.
   C) occurs with large amounts of detritis on the ocean floor.
   D) starts with kelp.
   Answer: B

Pages: 660

10. Succession in an intertidal area in Southern California started on barren rock with
    A) kelp.
    B) sea urchins.
    C) herbivorous fishes.
    D) diatoms.
    Answer: D

Pages: 661

11. The holistic concept views succession as
    A) population dynamics (competition, regeneration, mortality).
    B) random unorganized events.
    C) driven by changes in attributes between younger and mature stages.
    D) as reductionists view it.
    Answer: C

Pages: 663

12. Modern mechanistic succession models would include
    A) the resource ratio model.
    B) the individual-based animal model.
    C) the climax concept.
    D) the floristic relay.
    Answer: A

Pages: 664
13. The endpoint of succession is
    A) a sere.
    B) a pioneer community.
    C) an early seral stage.
    D) the climax.
    Answer: D

Pages: 663
14. At the start of a forest succession
    A) each community in the sere is a unique one time event.
    B) after a disturbance has occurred, the exact original community will not occur.
    C) climax vegetation over a region can be predicted but not local climaxes.
    D) all of the above
    Answer: D

Pages: 672
15. Stages of succession last
    A) one year.
    B) ten years.
    C) one thousand years.
    D) varying length of time; it depends on the sere.
    Answer: D

Pages: 664
16. The climax has _____ niches than a pioneer community.
    A) fewer
    B) more
    C) the same number of
    D) less dense
    Answer: B

Pages: 668
17. Fluctuations
    A) are short term reversible changes.
    B) have a floristic composition which is unstable.
    C) cause new species to appear.
    D) are dominant changes which are not reversable.
    Answer: A

Pages: 669
18. In the pioneer stage of forest succession
    A) soils are low in nutrients.
    B) organic matter is missing.
    C) sunlight is excessive.
    D) all of the above
    Answer: D

Pages: 669
19. A pioneer plant in the sand dunes would be
   A) an oak tree.
   B) a pine tree.
   C) shrubs.
   D) grasses.
   Answer: D

Pages: 672
20. With succession in plant life comes changes in animal life such as
   A) in the forest, grasshoppers become common.
   B) in the grasslands which become hosts to towhees, catbirds, and goldfinches.
   C) in the forest where woodpeckers appear as do the warblers and squirrels.
   D) all of the above
   Answer: C

Pages: 677
21. During the pre-Pleistocene at the beginning of the Miocene epoch
   A) tropical and subtropical forests existed farther north than today.
   B) Neotropical and Paleotropical forests covered most of central North America.
   C) Arcto-Tertiary forests of broadleaf and conifers extended across the northern U.S.
   D) all of the above
   Answer: D

Pages: 679
22. In North America, four great glaciers covered much of the northern part and
   A) they occured during the Jurassic period.
   B) afterward grasslands appeared over most of the eastern area
   C) buffalos became extinct.
   D) forests were reestablished in the eastern glaciated region.
   Answer: D

Pages: 664
23. The climax community theoretically is characterized by
   A) gross production = respiration.
   B) energy used = energy released by decomposition.
   C) a uniform climate.
   D) A and B.
   Answer: D

Pages: 659
24. Which of the following initiates succession:
   A) strip mining
   B) farming
   C) lumbering
   D) all of the above
   Answer: D

True-False

Write T if the statement is true and F if the statement is false.

Pages: 656
1. Seasonal species changes are called succession.
   Answer: False

Pages: 656
2. Bare rock will be colonized first by shrubs.
   Answer: False

Pages: 659
3. Colonizing organisms are K-selecting.
   Answer: False

Pages: 659
4. The plants in early stages of succession are fast growing and prolific seed producers.
   Answer: True

Pages: 659
5. According to the facilitation model organisms in one stage alter the environment so later stages can develop.
   Answer: True

Pages: 672
6. Animals do not correspond to plant communities during succession.
   Answer: False

Pages: 659
7. Secondary succession is usually initiated by the growth annual plants.
   Answer: True

Pages: 660
8. Pond succession does not include marsh plants.
   Answer: False

Pages: 660
9. Predation can have a major influence on successional development of kelp when it is preyed upon by sea urchins.
   Answer: True

Pages: 661
10. A community can be thought of as consisting of individual species responding individually to the environment.
    Answer: True

Pages: 664
11. According to the climatic climax theory all seral communitites will eventually converge to stabilize one climax determined by climate.
    Answer: True

Pages: 665
12. The pattern climax theory recognizes a variety of climaxes dependent on biotic and abiotic conditions.
Answer: False

Pages: 668
13. During fluctuations no new species are introduced.
Answer: True

Pages: 666
14. Cyclic succession is a replacement of climax species by non-climax species.
Answer: False

Pages: 672
15. Although it would be possible to predict that an old field in Appalachia will return to a forest by succession, the composition of that forest cannot be predicted.
Answer: True

Pages: 672
16. Each successional community occurs many times in a comparable area.
Answer: True

Pages: 672
17. Grasshoppers, sparrows, and many other animals are restricted to a successional grassland community in a forested region.
Answer: True

Pages: 676
18. Microcommunities lack microsuccession.
Answer: False

Pages: 677
19. Climatic changes of pre-Pleistocene and Pleistocene epoch are present today.
Answer: True

Pages: 679
20. The four glaciers of the Pleistocene and recent times determined the soils and topography of Canada and the middle of the northern part of the U.S.
Answer: False

Short Answer

Write the word or phrase that best completes each statement or answers the question.

Pages: 656
1. _____ is a sequential and directional change in species composition of natural communitites.
Answer: Ecological succession

Pages: 656
2. The actual sequence of communities in a succession from one start to a climax community is a/an _____ and each specific community a/an _____.
Answer: sere, seral stage (in order)

Pages: 656
3. The final seral stage in succession is the _____.
   Answer: climax community

Pages: 656
4. A succession starting in a bare area where no previous community existed is called a/an

   _____.
   Answer: primary succession

Pages: 656
5. _____ starts in an area previously occupied by some type of community. One example is old field succession on abondoned farms.
   Answer: Secondary succession

Pages: 656
6. Organisms which colonize barren areas in a primary succession are _____.
   Answer: pioneer species

Pages: 659
7. In _____, the pioneer species ameliorate the environment allowing other species to colonize the area.
   Answer: primary succession

Pages: 659
8. In secondary succession, initial plants grow fast, produce many seeds, and are a/an _____ species.
   Answer: r-selected

Pages: 660
9. Succession in aquatic environments starts with water and is therefore a _____.
   Answer: primary succession

Pages: 660
10. Intertidal recolonizations are initiated by _____.
    Answer: diatoms

Pages: 661
11. Clement's theory of succession begins with a/an _____, and finally ends with a/an _____ community which is stable.
    Answer: disturbance, climax   (in order)

Pages: 661
12. The _____ model of Egler is similar to Clement's as is the _____.
    Answer: floristic relay, holistic concept   (in order)

Pages: 661
13. The _____ is that competition and other population dynamics result in succession.
    Answer: reductionist concept

Pages: 664
14. The _____ theory states that there is only one climax and it is determined by climate.
    Answer: climate climax

15. The _____ theory states that the climax of an area is composed of a mosaic of vegetation climaxes controlled by soil moisture, nutrients, topography, slope exposure, fire, and animal activity.
Answer: polyclimax

16. _____ involves a repetition in sequence of seral stages brought about by the imposition of some disturbance.
Answer: Cyclic succession

Matching

Choose the item from Column 2 that best matches each item in Column 1.

1. Column 1: Later stage species not inhibited by early stage species.
   Column 2: Tolerance model of succession

2. Column 1: Each succession stage adds organic matter changing the environment
   Column 2: Facilitation model of succession

3. Column 1: Plants colonize an area keeping out other plants
   Column 2: Inhibition model of succession

4. Column 1: Plants nearer late stages of succession are more efficient in exploiting resources.
   Column 2: Tolerance model of succession

5. Column 1: Pioneer plants alter the environment
   Column 2: Facilitation model of succession

Essay

Write your answer in the space provided or on a separate sheet of paper.

1. What is the difference between primary and secondary succession? Why should they be considered as different?
   Answer: Primary succession starts on areas where no life previously existed. Pioneer communities developed and the sere for that situation started. Bare rock and scraped off areas are typical primary succession areas. Secondary succession starts in areas in which there have been previous organisms. The seral sequences of later stages in either may be the same.

2. What is a climax community? What factor determines the climax and what characteristics does it have?

    Answer: The climax is a stable end community of a successional sequence or sere. Theoretically, it is capable of self-perpetuation under prevailing environmental conditions. Because environmental conditions are variable, rarely do communities achieve equilibrium with the environment.

3. What would be the seral changes from bare rock to a deciduous forest?

    Answer: Successful pioneers have characteristics of fast growth: small size, short life cycle, and wide dispersal range. A common pioneer plant on bare rock is the lichen. After organic matter is life built, other grasses and wild forbs colonize the site followed by shrubs and small trees. Finally the forest develops. Soil is now well developed.

4. What are microcommunities? Do seres occur here?

    Answer: Microcommunities are small communities which differ significantly from the main community. Some microcommunities are rotten logs, patches of cow manure, or other small areas with their own characteristics. Microcommunities have micro succession. For example an acorn: invaded while still on the tree it is consumed and rotted by other organisms. In this case a parade of organisms and seral stages occurs until the microcommunity is completely broken down.

5. What are mechanistic models compared to other models? What is their importance?

    Answer: Mechanistic models are based on recent experiments and computer simulations. They consider competition, life histories, competition, and especially limited nutrients to explain succession. Other methods consider recruitment, features of the ecosystem.